84 68
WITHDRA
TELFREN
AF745145

UNIVERSITY COLLEGE OF WALES
LIBRARY
ABERYSTWYTH

# MODERN DATA COMMUNICATION

## Concepts, Language, and Media

# MODERN DATA COMMUNICATION

## *Concepts, Language, and Media*

WILLIAM P. DAVENPORT

*Vice President and General Manager*
*Scott-Buttner Communications*

Pitman Publishing

*To Maryon, my wife*
*without whose loving understanding*
*and infinite patience this book would*
*never have been completed*

FIRST PUBLISHED IN UK 1972

Published in the United States of America
by the Hayden Book Company Inc

© Hayden Book Company Inc 1971

SIR ISAAC PITMAN AND SONS LTD
Pitman House, Parker Street, Kingsway, London WC2B 5PB
PO Box 46038, Portal Street, Nairobi, Kenya
SIR ISAAC PITMAN (AUST) PTY LTD
Pitman House, 158 Bouverie Street, Carlton, Victoria 3053, Australia
PITMAN PUBLISHING COMPANY SA LTD
PO Box No 11231, Johannesburg, South Africa
PITMAN PUBLISHING CORPORATION
6 East 43rd Street, New York, NY 10017, USA
SIR ISAAC PITMAN (CANADA) LTD
495 Wellington Street West, Toronto 135, Canada
THE COPP CLARK PUBLISHING COMPANY
517 Wellington Street West, Toronto 135, Canada

Cased edition ISBN: 0 273 25215 1
Limp edition ISBN: 0 273 25218 6

Printed in Great Britain by Unwin Brothers Limited
The Gresham Press, Old Woking, Surrey, England
A member of the Staples Printing Group

G2(T1267/1367:74)

THE UNIVERSITY COLLEGE OF WALES
ABERYSTWYTH
ACC. No. B74-9417
CLASS No. TK 5102.5 D2

# Preface

Data communication may be defined as the movement of information in machine language. The term usually connotes a system in which business machines exchange information by means of an interconnecting electrical communication link. The business machine could be a teletypewriter, card reader or punch, paper or magnetic tape terminal, computer, or a wide variety of other devices. The kinds of data moved could include information from source documents, computer programs, business record files, keystrokes from a direct-inquiry device, handwriting, or almost any other kind of information.

Data processing has revolutionized backstage business operations. Data communication is already beginning to stretch the minds and talents of many forward-looking men. For example, H.I. Romnes, president of American Telephone and Telegraph Company, said at the Chicago International Data Conference in 1966, "... while data processing and data transmission are separate undertakings, each must take increasing account of the other if the full potentialities of both are to be fulfilled."

Richard Bloch, a Honeywell vice president, once said, "The computer that lacks ability to communicate with the outside world is in danger of remaining an isolated marvel, mumbling to itself in the air-conditioned seclusion of its company's data processing room." C.W. Spangle, EDP vice president of the Business Equipment Manufacturers Association, emphasized the extent of the job ahead of us when he spoke at the BEMA Convention in Chicago in November 1968: "We project growth on the magnitude of 10 to 15 percent a year in the overall data processing market, with certain segments growing somewhat faster. Foremost among them, I believe, is communications. Approximately 12 percent of the computers installed around the world today are operating in a communications environment. But about 21 percent of the computers shipped this year will be communications oriented. . . . You can see from these statistics that communications . . . moving data from person to machine and back or from machine to machine and back . . . *will be the big business for the computer industry in the years ahead.*"

Although the possibility of data communication is no longer a dream, relatively few people realize how data is transmitted or how to evaluate the benefits of data communication. This book is written to explain the theories and concepts of data transmission to the operators and managers of information systems, and the students of information handling. An attempt has been made to reduce intricate theories to clear and understandable terms without losing the essence of their meanings. To this end, only simple arithmetic is used, and all terms, calculations, and assumptions have been explained in some detail.

ABERYSTWYTH
LIBRARY
U.C.W.

Data communication, like so many other technical subjects, is actually an assembly of other technologies, each of which must be independently developed. Although the chapters are arranged in the approximate order of complexity of their contents, it would be possible to study them in any order.

Virtually all of the topics covered in this book would be applied during the design and installation of a data communication system. The contents, therefore, have been arranged to serve as a reference or handbook as well as an introductory text. Frequent use of the glossary will provide clarification and elaboration of text material.

**Alameda, Calif.** **William P. Davenport**

# Contents

*CHAPTER 1*

# *Introduction to Data Transmission*

The term "data" as used in this book refers to information that may have been taken from source documents, such as sales orders, time cards, production records, and the like; from some storage medium, such as magnetic tape, punched cards, or paper tape; or from the memory of a computer. The movement of this data over a distance between machines is data transmission. The machines involved in the communication of data range from a simple paper-tape reader to a complex computer, and their languages are a variety of codes (usually binary) that can be interpreted directly by the machine.

## Background

Decentralization was the trend of many businesses and industries during the 1950s. Although it seemed to solve many operational problems, it revealed and intensified the problems associated with communications. More copies of documents, reports, and source data had to be produced. Machine installations duplicating those already in existence had to be provided at outlying points. The transportation of volumes of information became too slow and too expensive. Operations and decisions were detained while information squeezed through the communications bottleneck.

## Requirements, Past and Present

As a result of this pressure from business and industry, imaginative business-machine manufacturers began to look for economical methods of moving and handling the tides of paper and messages moving between the places they were generated and their destinations. The requirements of the proposed transmission system were the same as they are today: accuracy, simplicity, speed, flexibility, and economy. A system has to be simple enough to be operated with almost no training involved and to require no major changes to transmission terminal equipment and facilities. The system has to be fast enough to meet the requirements of the users of the information and to provide a margin of desirability over alternate transportation methods. It cannot be so rigid in its configuration, geography, or both to prohibit changes of application or terminal relocations. It has to compete favorably on a dollar basis with alternate methods and maintain a high degree of accuracy. This new concept of data transmission had to be as comprehensive as possible; it could not be limited to one speed, one kind of machine or terminal, or one application.

Studies by Bell Laboratories and others revealed that if electrical pulses generated by business machines were converted to audible tones approximating the range of the human voice, these tones could be carried by the same network and equipment that was used for ordinary telephone conversations. Devices called "data sets" were developed to perform the conversion function, and machines began "talking" with other machines over the telephone network. Thus evolved the concept of data transmission — data (information in machine language) transmitted over existing voice telephone lines, with provisions made for voice communications on an alternate basis using the same data equipment if required.

**Areas of Use**

The areas to which data communications may be profitably applied have become so broad and numerous that it is safe to describe them in generalities. Nearly every category of business machines can be adapted to data transmission: adding machines, typewriters, keypunches, cash registers, gasoline pumps, computers, and even the lowly pencil! Some have been arranged to handle data stored in a medium, such as paper tape or punched cards, which act as a buffer or storage until the information is transmitted by another machine; other machines connect directly to transmission facilities. It was once thought that a limited variety of data sets would provide the diversity necessary to accommodate most business machine transmission requirements, but the field has evolved to more than 50 different sets in the Bell System alone. Those produced by Western Union and private industry add to this figure appreciably. With this variety of data sets and the associated business machines, present technology can solve most information transportation problems.

Which problems can be solved or simplified by data communications? Business operations that have any of the following symptoms might benefit from the use of data communications equipment or services:

(1) Decentralized operations
(2) High volume of mail, messenger service, or phone calls between locations
(3) Manual checking or duplicate transmission to assure accuracy

(4) Repetitive paperwork
 (a) Duplication on successive forms
 (b) Re-creation or copying
 (c) Large number of forms
 (d) Same information on different forms
(5) Many standard typewriters used in information flow process
(6) Use of punched card or computing equipment
(7) Lack of timely control and updating information
(8) Slow order processing, billing, or shipping
(9) Too much money in inventory

To help crystallize facts, figures, and thoughts concerning possible requirements for data communications, the following questions concerning the seven criteria of data flow should be considered:

(1) *Function:* Who needs the information? Why? What does the message call for — action? information?
(2) *Distribution:* Where does the message originate? Does the information flow from one point to another, one point to many, many to one? How are messages relayed?
(3) *Volume:* What amounts of all kinds of information (phone calls, forms, teletypewriter, data, etc.) are moved within a given period of time (usually an hour or a day)?
(4) *Urgency:* How soon after origination must the message arrive — immediately? within an hour? within a day?
(5) *Language:* What is the physical nature of the message — voice? handwritten? typewritten? machine language?
(6) *Accuracy:* How many and what kinds of errors can be tolerated — transmission? equipment? human? What accuracy is needed?
(7) *Cost:* How much are you prepared to pay for improved communication facilities? At what point does cost begin to exceed value received?

## TRANSMISSION SYSTEM COMPONENTS

The three basic components common to all communication systems are the *source,* which originates the information; the *medium,* through which the information flows; and the *sink,* which accepts the information (see Fig. 1-1). An example could be a private-line teletypewriter system. The transmitting teletypewriter (source) creates direct-current (d-c) pulses (signals) which are sent over wires (the medium) to a receiving teletypewriter (sink), which converts the signal back to its original form in characters.

Figure 1-1. *The three primary components of a data transmission system.*

### The Source

In the above example, the teletypewriter, acting as the source machine, received information from keyboard depressions. As each key was depressed, the machine encoded the corresponding character into a discrete combination of direct-current pulses for transmission. The function of any source machine is to encode and transmit the information offered to it. In some cases the information will already be in code form in paper or magnetic tape, punched cards, computer memory, and the like. If so, the source machine must interpret and transform the coded information into corresponding electrical pulses.

### The Medium

The medium carries the transmitted information from the source to the sink. Examples of transmission media include wire, radio, coaxial cable, microwave, and light beams. Since characteristics of these media differ widely, information must be prepared in a particular way for each. To travel over wires between telephones, for example, voice must be transformed into electrical energy. The direct current of a business machine must be converted to alternating current before it can be transmitted over the switched telephone network. Media such as radio and coaxial cable require more sophisticated preparation of the signals.

### The Sink

The sink is the recipient of the transmitted information. In many cases, the sink will be the same type of machine as that of the source. The teletypewriter system described above was such an example. In many other instances, the sink will differ considerably from the source machine. For example, it is becoming commonplace for teletypewriters and display terminals (using cathode-ray tubes and keyboards) to converse directly with computers. In any case, the sink is the machine receiving information at any given time. In two-way communication systems, one terminal may alternate in its role as source and sink.

### The Modulator/Demodulator

The source, medium, and sink are common to all communication systems. As such systems become more sophisticated, additional factors may have to be considered (see Fig. 1-2). An example above discussed teletypewriters which exchanged information in the form of d-c pulses over a private wire. If the same machines were to transmit over the public telephone network, the d-c signals would first have to be converted to various frequencies, or tones. This conversion function is called modulation, and it is performed by a modulator. At the receiving end of the system a device called a demodulator converts the tones back to the original d-c

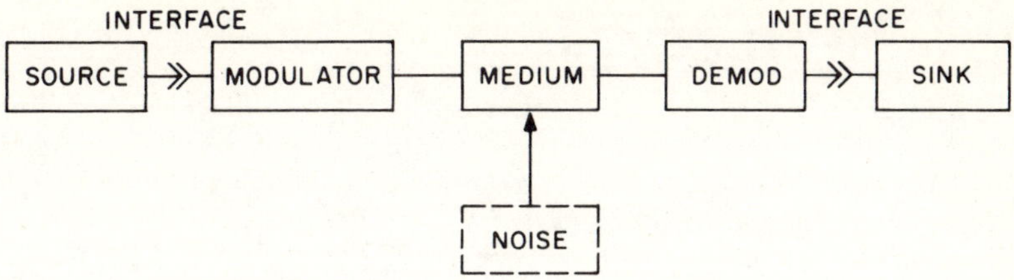


Figure 1-2. *Data transmission system block diagram.*

pulses, which are the native language of the receiving machine. These modulators and demodulators are often called "data sets" and will be discussed in detail in a later chapter.

### The Interface

In most data transmission systems, between the business machine and its associated modulator or demodulator, is a plug which serves to connect the electrical circuitry of the units and to act as their mutual boundary. This plug is called an *interface,* and its characteristics are standardized through regulation by Electronics Industry Association (EIA) standards.

### Noise

Although noise (unwanted electrical impulses) is not a part of the physical components of a transmission system, its presence is real. It is most generally associated with the transmission lines, through which it normally gains access to the system. The system's components must be designed to overcome a certain level of noise. Full treatment of the subject of noise is given in a later chapter.

## DIVISION OF RESPONSIBILITIES

### Business-Machine Manufacturers

A business machine acts as the source or the sink in a transmission system. The machine's main function is often not that of transmission: cash registers, gasoline pumps, computers, and the like. But if the business machine is to be used for transmission of data, the machine must conform to certain standards and procedures. It is the responsibility of the machine manufacturers (or their representatives) to maintain the *machine functions* of the system and to assure that a signal that conforms to predescribed characteristics is provided at the interface.

Input to a business machine may consist of keystrokes, punched cards, magnetic or paper tape, telemetry signals, or even handwriting. The function of the business machine, if it is to be used as a transmission device, is to transform these media into a signal that can be put directly on a circuit or first given to a data set to condition for line use. The receiving business machine takes the signals from the receiving data set and converts them back to a machine media (paper or magnetic tape, punched cards, etc.) or produces a printed page, a video readout, handwriting, and so on, that can be interpreted directly by humans.

## Common Carriers

The responsibilities of the common carriers consist, in part, of the transmission functions of the system and thus lie between the interfaces. Common carriers are responsible for accepting a signal from a business machine, transporting that signal to a distant point, and delivering the signal exactly as it was received. The method used to transport the signal varies, but should be of no concern to the business-machine people or to the user as long as the signal conforms to specified standards. (Actually, the responsibilities of a common carrier go far beyond the interfaces — systems design and assurance of equipment compatibility, for example — but the above comments pertain to the transmission system alone.) During the transmission of the signal, the common carrier must maintain the signal's integrity by protecting it from distortion, fading, noise, and loss.

## The System User

The operational functions of the system are the responsibility of the user. He must conform to the procedures and formats required by the

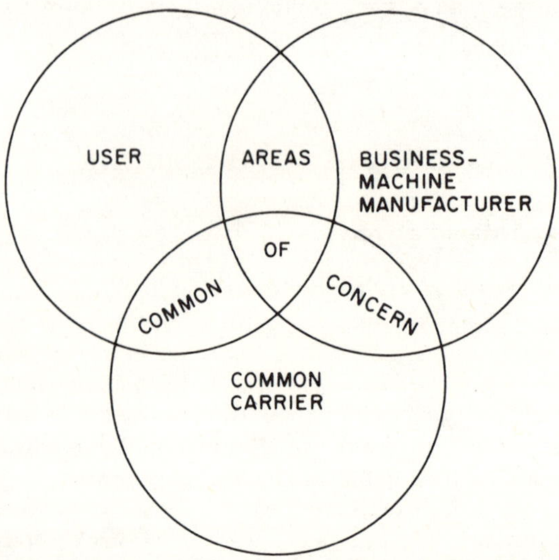


Figure 1-3. *Relationship of user, business-machine manufacturer, and common carrier.*

machine. He must also protect it from damage and pay for the power and help necessary to operate (and often to maintain) it. He is responsible for deciding the best uses for the system, and in doing so should call in all available outside help. The machine vendors are specialists in their fields and have the advantage of training and previous experience. The user should not become committed to any line of action until at least two machine vendors have given their suggestions. The common carriers also offer valuable advice and assistance that the user should utilize if he is considering a communications system. Ideally, meetings should be held with representatives from the common carrier, the machine vendor, and the user prior to and during an installation of any magnitude. These meetings are the responsibility of the user, since the final decision is his. (See Fig. 1-3.)

## Problems

1. Define data and data transmission.
2. List six symptoms of the need for data communications.
3. What are the seven criteria of data communications?
4. (a) What are five main components of a communications system, excluding noise? (b) What are the functions of each? (c) Give two examples of communications systems and name the components of each.
5. What are the prime responsibilities of the following concerning a data communication system: business-machine supplier, common carrier, user?

*CHAPTER 2*

# *The Language of Data*

## INTRODUCTION

The data transmission student must understand certain fundamentals concerning data: what it is, how it is measured, and how it differs from other kinds of information. This chapter has been divided into three parts. The first portion has been written for those who have no background in the theory of numbering systems. Numbering systems form the roots of the languages of data communication, and a knowledge of these simple fundamentals is essential to the understanding of code construction, error-control methods, and the operation of data sets.

The second portion discusses the value and measurement of information and defines and compares bits and bauds. The third portion introduces the concept of information coding.

## CONCEPTS OF NUMBERING SYSTEMS

### Place Value, Base, and Power

"Place value" refers to the relative value a digit assumes in various positions within a number. The digit 5 in the units position of the number 585 yields a relative value of five, whereas the 5 in the hundreds position yields a relative value of five hundred. The value of any number is thus found by multiplying each digit by its place value, and adding the result: $(5\cdot100) + (8\cdot10) + (5\cdot1) = 585$.

Place values of any numbering system are found by taking successive powers of the base of the system. The place values of the decimal system, therefore, are:

$10^0 = 1$
$10^1 = 10$
$10^2 = 100$
$10^3 = 1000$
etc.

The result of taking the zero power of any number is always one, and a number to the power of one is the number itself. When numbers of several bases are being discussed, confusion is lessened if the numbers are written with a subscript denoting the appropriate base. Decimal 13 would thus be written $13_{10}$, and octal thirteen would appear as $13_8$, etc.

Working with decimal numbers is so automatic for most people that the above concepts are used unconsciously. They were reviewed, however, because they hold for *any* numbering system.

## THE BINARY NUMBERING SYSTEM

The binary numbering system is the true language of data, since most electromechanical and electronic data devices operate in two states: on or off, plus or minus, and the like. A working knowledge of binary is required for developing an understanding of data communication.

### Comparison of Binary with Decimal

All numbering systems have several things in common:

(1) The base is indicated by the system's name (and vice versa), and equals the number of digits in the system.
(2) The value of any digit in a system will not exceed the value of the base minus 1.
(3) The 0 indicates that the end of the basic system has been reached.
(4) Successive powers of the base indicate the place values of the system.

Apply these rules to the binary system of numbering:

(1) The name binary indicates that two digits make up the system, and thus it has a base of 2.
(2) The highest value of any digit in the system is one less than the value of the base. Since this base is 2, the highest digit is 1. The other digit, of course, must be 0.
(3) Begin counting in binary, starting with 1. Since the base is 2, the next number must be the last in our basic counting system, and this situation is indicated by the use of the zero. Thus the second number, after 1, is 10.
(4) The place values of digits in a binary number are the successive powers of its base, 2:

$2^0 = 1$
$2^1 = 2$
$2^2 = 4$
$2^3 = 8$
$2^4 = 16$
etc.

## Converting Decimal to Binary

A decimal number is converted to binary by repeatedly dividing by the base and recording the remainders. For example, the decimal number 42 is converted to binary in Table 2-1. The binary equivalent of decimal 42,

TABLE 2-1. CONVERSION FROM DECIMAL TO BINARY

| Decimal Number | | Base 2 | | Quotient | Remainder (Binary) | |
|---|---|---|---|---|---|---|
| 42 | ÷ | 2 | = | 21 | 0 | |
| 21 | ÷ | 2 | = | 10 | 1 ↑ | |
| 10 | ÷ | 2 | = | 5 | 0 | |
| 5 | ÷ | 2 | = | 2 | 1 | |
| 2 | ÷ | 2 | = | 1 | 0 | (Read up) |
| 1 | ÷ | 2 | = | 0 | 1 | |

therefore, is 101010. To check the work, convert this back to decimal by one of the following methods.

## Converting Binary to Decimal

Two methods of conversion are described below. The first is demonstrated in Table 2-2. The descending powers of two (for a binary chart) are entered on the upper line, and the derived place values below them. The binary number is entered in its normal sequence toward the right (low-order) edge of the chart (right justified). The decimal equivalent is obtained by adding the place values appearing above each 1 in the binary number. In this case, place values of 32, 8, and 2 were added to give 42. This method can be used to find the decimal equivalent of a number in *any* base. Each place value is multiplied by the corresponding digits in the number below it, and the products are added to yield the decimal equivalent. Table 2-2 shows powers of two up to $2^8$. The chart could be extended to as high an order as desired; for example, Table 2-3 shows place values up to $2^{19}$.

TABLE 2-2. BINARY TO DECIMAL CONVERSION CHART

| | | | | | | | | | | |
|---|---|---|---|---|---|---|---|---|---|---|
| Powers of 2: | $2^8$ | $2^7$ | $2^6$ | $2^5$ | $2^4$ | $2^3$ | $2^2$ | $2^1$ | $2^0$ | |
| Place values: | 256 | 128 | 64 | 32 | 16 | 8 | 4 | 2 | 1 | |
| Binary number: | | | | 1 | 0 | 1 | 0 | 1 | 0 | |
| | | | | ↓ | | ↓ | | ↓ | | |
| | | | | 32 + | 0 + | 8 + | 0 + | 2 + | 0 = | 42 |

**TABLE 2-3. POWERS OF TWO**

| N | $2^N$ | N | $2^N$ |
|---|---|---|---|
| 0 | 1 | 10 | 1024 |
| 1 | 2 | 11 | 2048 |
| 2 | 4 | 12 | 4096 |
| 3 | 8 | 13 | 8192 |
| 4 | 16 | 14 | 16,384 |
| 5 | 32 | 15 | 32,768 |
| 6 | 64 | 16 | 65,536 |
| 7 | 128 | 17 | 131,072 |
| 8 | 256 | 18 | 262,144 |
| 9 | 512 | 19 | 524,288 |

## Double-Dabble

Another useful method of converting binary to decimal is "Double-Dabble," which permits mental conversions without resorting to a written table or chart. Double-Dabble has only three rules:

(1) Use the first binary 1 unchanged
(2) Double on "0" (Double-double)
(3) Dabble on "1" (Dabble-double and add 1).

This may appear confusing, but it is the simplest method of converting binary to decimal. Table 2-4 explains in more detail the three rules given above.

**TABLE 2-4. USING DOUBLE-DABBLE TO CONVERT FROM BINARY TO DECIMAL**

| Binary Digit (101100) | Action | Process | Decimal Result |
|---|---|---|---|
| 1 | Carry | | = 1 |
| 0 | Double | 2 x 1 | = 2 |
| 1 | Dabble | (2 x 2) + 1 | = 5 |
| 1 | Dabble | (2 x 5) + 1 | = 11 |
| 0 | Double | 2 x 11 | = 22 |
| 0 | Double | 2 x 22 | = 44 |

Convert binary 101100 to decimal using Double-Dabble. Rule 1 says don't Double or Dabble the first binary one, so carry it as the digit one. The next digit is a zero, so Rule 2 applies. It says a zero means to double the carry, so double the 1. The next binary digit is a 1, so follow Rule 3; double the number and add one to it. The value of two is being carried from the previous calculation, so double it for four, and add one for five. The next 1 again means Dabble, so five doubled is 10 and one is 11. The next two digits are zeros, so double 11 twice and get 44 as the decimal equivalent of binary 101100.

### Binary Addition

The rules of binary addition are identical to those for decimal addition. The process of binary addition, however, is even more simple, since there are only two digits to work with.

*Rule 1:* $0 + 0 = 0$
*Rule 2:* $0 + 1 = 1$
*Rule 3:* $1 + 0 = 1$
*Rule 4:* $1 + 1 = 10$

When two binary ones are added, the one in the sum is carried, just as in the decimal system. Adding a series of binary ones is illustrated:

| | | | | | | | | |
|---|---|---|---|---|---|---|---|---|
| Carries: | | 11 | | 1 | | 111 | | |
| 1 | 10 | 11 | 100 | 101 | 110 | 111 | 1000 | 1001 |
| +1 | +1 | +1 | +1 | +1 | +1 | +1 | +1 | +1 |
| 10 | 11 | 100 | 101 | 110 | 111 | 1000 | 1001 | 1010 |

A process often used in error-control systems is binary "half-adding." This term merely means that any carries resulting from binary addition are dropped, leaving only one significant binary digit. If A and B are binary digits, C is the result of half-adding as shown in Table 2-5.

TABLE 2-5.

| INPUT | | | | OUTPUT |
|---|---|---|---|---|
| A | + | B | = | C |
| 0 | | 0 | | 0 |
| 0 | | 1 | | 1 |
| 1 | | 0 | | 1 |
| 1 | | 1 | | 0 |

## DEFINING INFORMATION

### Value of Information

Most people have always taken the word *information* for granted, as if it needed no definition. They have given and taken information, processed information, and have felt the need for information. A little thought, though, will reveal that information highly prized by some people means little or nothing to some others.

News that Gigo Computers, Inc., underbid all competitors for a contract with the federal government, for example, would obviously have more meaning to a particular group of people than it would to others. The news of the winning bid carried the same *quantity* of information to everyone that heard it, but it had a different *significance* in each case. The term information, then, lacks any degree of value except that which is assigned by the recipient. Instead, *information is an inherently quantitative term and is measured by the degree to which it clarifies that which is unknown;* a totally predictable event contains no information. It is the *quantitative* value of information (as opposed to its emotional value) that will be discussed in the balance of this book.

The paragraphs below will reveal exactly how much information was contained in the news that Gigo had won the contract. An understanding of this concept will be helpful during the study of efficiency, coding schemes, and error control, all of which are discussed in following chapters.

## Value of a Message

A *message,* in information theory, is merely the output of some information source. If the source were a telephone transmitter, the message would be the analog voltages impressed on the line. If the source were a teletypewriter, the message could be a character, one of the bits making up a character, or an entire word. The composition of a message, then, can vary, and it must be defined, or understood, by its use in a system.

The quantitative value of a message is based on several factors. First, it must be established how much was known about the contents of the message before it was received. If it was known that Gigo was the only company bidding, the news that it had won would have been no surprise, and the information content of the message would have been zero. A message giving the value of the signed contract, however, would contain a certain amount of information; Gigo was known to be bidding, but no one knew exactly what amount the bid specified.

Second, to further define the quantity of information a message carries, it must be known how many messages were in the set from which this message was chosen. If three companies bid for the contract, one of three possible messages will have to be sent to identify the winner. If ten companies bid, the winning message must be chosen from a group of ten and would have to carry more information to identify the winner.

Third, to be very precise, one would also have to know the probability of each event that the message could describe. If each of the ten companies bidding had exactly equal sales records, the message naming the winner would carry everything to be learned about the outcome of the bid. If the sales records were different for each company, though, one would have some information about the outcome before the results were known, and thus the winning message would carry that much less information.

Calculation of the information content of messages of unequal probability becomes quite complex, and is beyond the scope of this book. To help clarify this discussion, therefore, assume that information is sent in random form, and thus that all messages are equiprobable.

### Entropy

In general, it can be said that information has the property of reducing the uncertainty of a situation. The uncertainty is called *entropy* (H), and entropy exists to the extent that information is lacking (Information + Entropy = 100%, or, in short form, I + H = 1). If the entropy of a situation is small, then only a small amount of information is required to clarify it. If the entropy is large, then much more information will be required before the uncertainty is replaced by an acceptable degree of clarity.

If Gigo had bid against only Bells and Whistles, Inc., there would have been a small amount of entropy; only two manufacturers were bidding and both had the same sales records. If Gigo had bid against a group of ten competitors, however, the entropy would have been large, because not only would there have been a greater number of companies, but each would have had a different probability of winning the contract.

Some transmission codes are designed to reduce entropy to the point where errors incurred during transmission can be not only detected and located, but even corrected. This subject will be discussed in a later chapter.

## BITS

### Definition of Bit

A binary condition is one having a bistable quality. Thus, either of two conditions may exist: on or off, yes or no, mark or space, magnetized or demagnetized, and the like. In the binary numbering system, these two conditions are represented by the digits 1 and 0. It was inevitable that someone would abbreviate "binary digit"; the result was the term "bit."

Note that *both* binary states are called bits, not just the 1 bit. The reason is that they both carry the same amount of information; the presence of one means the absence of the other. Compare this with the decimal system: the presence of, say, the numeral 8 has a definite meaning, but the mere absence of 8 could have a variety of meanings!

### Measuring Information

Now that it has been established that information is a quantitative term, it should be determined how to measure it exactly. When Gigo bid against Bells and Whistles, there was a choice between two messages to designate the winner. It happens that a fundamental property of a binary bit (either 1 or 0) is the capability to *halve the uncertainty of a situation.* Since only two companies were bidding on this contract, the binary 1 could have been used to mean that Gigo had won, and the 0 for Bells and Whistles' victory. Thus the news that Gigo won carried one bit of information.

For comparison, how much information, or how many bits, are required to indicate the winner of a contract bid won by eight companies? If the companies are listed in order, and it is specified that a 1 bit means the upper half of the list and a 0 bit means the lower half, a series of bits can then be sent to indicate the winning bid. Since the winner, Gigo, is in the lower half of the list, the first bit will be a zero:

| | |
|---|---|
| Eliminated | Big Brother Mfg. Co.<br>Do-All Data<br>Itty Bitty Machines<br>Bells and Whistles, Inc. |
| 0 = | Digilog Corp.<br>Gigo Computers, Inc.<br>Picocycle Corp.<br>Memory, Unlimited |

Of the remaining half of the list after the upper portion has been eliminated, Gigo is in the upper half so a 1 is sent:

| | | |
|---|---|---|
| | | Big Brother Mfg. Co.<br>Do-All Data<br>Itty Bitty Machines<br>Bells and Whistles, Inc. |
| 0 | 1 | Digilog Corp.<br>0→ Gigo Computers, Inc. |
| | | Picocycle Corp.<br>Memory, Unlimited |

The next bit was a zero, since Gigo is the lower of the two remaining companies. Three bits, then, (010) have been used to define one of eight possibilities. No matter which bidder had won, only three bits would have been required to indicate the winner. Basically, a series of questions has been asked (upper half?) which resulted in "yes" or "no" answers. This was an example of a primary property of a bit; it not only is the smallest particle of information, but it also is the maximum amount of information possible in a yes or no selection.

**How Many Bits?**

With this knowledge, it can be determined how many bits are required to define one of any given number of possible selections. Remembering that a bit can be either a one or a zero, one bit can be used to define either of two selections, two bits to define one out of four selections, three bits for one of eight selections, and so on. Every added bit *doubles* the number of possible selections; the selections are increased in powers of two.

One bit: $2^1 = 2$ selections
Two bits: $2^2 = 4$ selections
Three bits: $2^3 = 8$ selections

.

.

.

etc.

If the number of possible selections is known, it can be determined how many bits will be required to identify any *one* of them. If there are eight companies bidding, then $8 = 2^3$, so three bits are required, as seen in the previous example. If only four companies had bid, then $4 = 2^2$, and two bits would have done the job. What if six companies had participated in the bid? Six is less than $2^3$, but greater than $2^2$, so a minimum of three bits would be needed to identify the winner.

Concerning the formula $2^3 = 8$, the 8 was known (the number of companies bidding) and the 2 was known (with a binary situation), but we wanted to find the unknown value of 3 (the minimum number of bits to identify one of eight companies). Working with the known values (8 and 2), the problem could have been approached by finding the logarithm of 8 to the base 2 (written $\log_2 8$). Another word for logarithm is exponent, so we are actually looking for the exponent of 2 that will give 8. Since $2^3 = 8$, then $\log_2 8 = 3$.

The number of bits required to identify any particular selection from a group of N possible selections is $\log_2 N$, provided that all N selections have equal opportunity of being chosen:

$$I = \log_2 N$$

A common communication code contains 32 different characters. How many bits must it have per character? Answer: $\log_2 32 = 5$ bits. ($2^5 = 32$). How many bits are required for a code which can describe all 26 characters of the alphabet without duplication? Answer: $\log_2 16 < \log_2 26 < \log_2 32$, so $\log_2 32$, or 5 bits, are required. If a question has only one possible answer, the information content of that answer will be zero: I (information) $= \log_2 1 = 0$. If a question had eight possible answers, as in the above race results, then the answer contains $I = \log_2 8 = 3$ bits.

The following example can serve to further clarify this concept.[1] The number of different ways that the cards in a full deck can be arranged is 52! (52! is read "52 factorial," and means 52·51·50. . . . .3·2·1). Using our general formula, we find that the number of bits that can be represented by the order of the cards after any particular shuffle (their order could be a "message") is: $I = \log_2 52! = 225.7$ bits. During any one shuffle, the deck is approximately halved into two packs, A and B. The A and B portions present

[1]Gordon Raisbeck, *Information Theory*, Cambridge, Mass.: The MIT Press, 1964.

a binary situation, and there are 52 cards in the total deck, so A and B can be combined in $2^{52}$ possible sequences. We can use the information formula to find the maximum number of bits that one shuffle can produce: $I = \log_2 2^{52} = 52$ bits. ($\log_2 2^{52} = 52 \log_2 2 = 52 \cdot 1 = 52$). Then 225.7 divided by 52 equals 4.3, so a minimum of five shuffles is required to assure that all 52 cards are in random order. Remember this during your next poker game!

### Bits and Other Bits

It might be assumed from the above discussion that all bits carry information. This is not so, and a clear distinction should be made between information bits and those that carry no information. The difficulty of making this point clear in the following discussion is compounded by the lack of names or terms for the two kinds of bits. G.C. Tarr of England has suggested the term "shannon" for the information bit, which is used to designate a choice between two equiprobable events. Perhaps this term will find acceptance. In this book, however, we will try to avoid confusion by using adjectives such as "information," "parity," "timing," and so forth to describe the particular bits being discussed.

A binary digit may or may not carry information. A rule of thumb, which works most of the time, is that if the value of the bit (1 or 0) cannot be predicted accurately, it contains information. If its value can be predicted accurately, it is serving a practical function other than carrying information. Information bits, then, are used in various combinations to create the codes used to stand for letters, numerals, and special functions; noninformation bits are used to permit the system hardware to tell where one character stops and the next one begins, to provide a means to detect and correct errors, and to permit equipment at opposite ends of a circuit to achieve and maintain synchronization. This distinction should be kept in mind, since it will clarify later discussions concerning coding, error control, and system efficiency.

## BAUDS

### Definition of Baud

In very much the way a bit is a unit of information, a *baud* is the *unit of signaling speed.* It is characterized by the presence or absence of a pulse in a channel and is a measure of the maximum rate of pulses (code elements) per second in a system. Technically, the baud rate is found by taking the reciprocal of the length (in seconds) of the shortest pulse used in creating a character. For example, the length of a pulse of baudot code as used with 60 wpm Bell System teletypewriters is 0.022 seconds (22 milliseconds, or ms). The baud rate is the reciprocal: $1/0.022 = 45.45$ bauds, which is the highest possible baud rate of this signal. See Fig. 2-1.

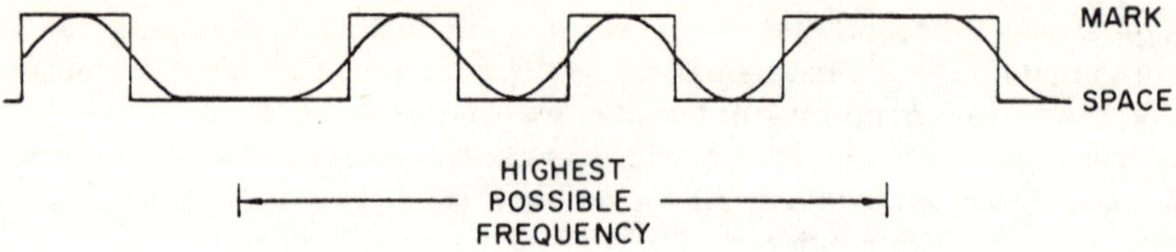


Figure 2-1. *Highest possible signal frequency.*

**Bauds vs Bits**

A long and heated controversy continues on the relationship of bauds and bits. The most common error is the use of bit and baud as synonyms. There is a clear distinction between a bit as a unit of information or a binary number and a baud as a unit of signaling speed. In fact, the bit rate and the baud rate of a signal rarely coincide. It is important to clarify this argument, because an understanding of the differences between bits and bauds will be of great assistance in the comparison of the true "throughput," and thus the efficiency, of various types of communications equipment. One type may appear attractive due to its declared high "bit rate," while it actually may be very inefficient and costly in comparison to similar types.

The argument generally centers on the output signal produced by a machine. In many cases, a business machine will generate some sort of timing, parity pulse, or both in addition to the original coded data pulses. In these cases, the baud rate will exceed the bit rate; the timing or parity bits merely add redundancy to the coded character and do not change the information content of the character.

The *Lenkurt Demodulator* (September, 1966) dealt with data communications and contained this exceptionally cogent comparison of bits and bauds:

> .... The speed of signaling, measured in terms of the amount of information transmitted per unit time, depends on the transmission path and its associated apparatus. Bits per second expresses the total number of *information* pulses in one second and includes redundant bits used for checking errors. If the pulses are of varying length, or if start and stop pulses between each character are added that are not part of the message, the bit rate tells nothing of their number or duration. On the other hand, baud — from Jean Maurice Emile Baudot, an officer in the French Telegraph Service, who contributed to early telegraph principles — is defined as the reciprocal of the time of the shortest signal element in a character. The term baud is often misinterpreted as a synonym for bits per second. However, the number of bauds equals the number of bits per second only when all time intervals are constant, and all signal pulses are information pulses, such as in binary transmission.
>
> An example of the relationship between bits and bauds is in

ordinary (100 wpm) teletypewriter transmission, which makes use of a five-bit code, each bit being 13.5 milliseconds in length. The baud rate is therefore the reciprocal of 13.5 milliseconds, or approximately 74.2 bauds. A single character consists of a start pulse and the five information pulses or bits, each of 13.5 milliseconds duration, for a total of 81 milliseconds. A stop pulse of 19 milliseconds ends the character. The total time for a character is then 100 milliseconds.

Since the bit speed depends on the number of *information* pulses transmitted per unit of time, the equivalent rate for this type of transmission is 5/100 ms, or 50 bits per second.

Now, if a lapse period of 20 milliseconds is arbitrarily inserted between the stop pulse of this character and the start pulse of the next, the bit rate would be reduced to 5/(100 ms + 20 ms) or 41.7 bits per second. However, the teletype speed would remain at 74.2 bauds, because the baud rate depends only on the time length of the shortest pulse (13.5 milliseconds) in the character.

The number of words per minute can be determined using the ordinary telegraph definition of a word, which is 6 characters. The speed in bits per second is converted into bits per minute by multiplying by 60; hence, 50 bits per second equals 3000 bits per minute. Since there are five bits per character and six characters per word, there is a total of 30 bits per word. Dividing 3000 bits per minute by 30 bits per word equals 100 words per minute. Here again, the transmission rate in words per minute could be reduced by a slow teletypewriter operator, but the signal speed remains at 74.2 bauds.

It can be concluded that the baud rate is very important to the telephone engineer, since this rate establishes the type of telecommunications channel to be used. The computer engineer is concerned to a lesser degree with baud rate, but economics and speed of information flow are uppermost in (his) technology. Hence, to him the bit rate is the major concern, and is the more common expression used in dealing with binary data transmission.

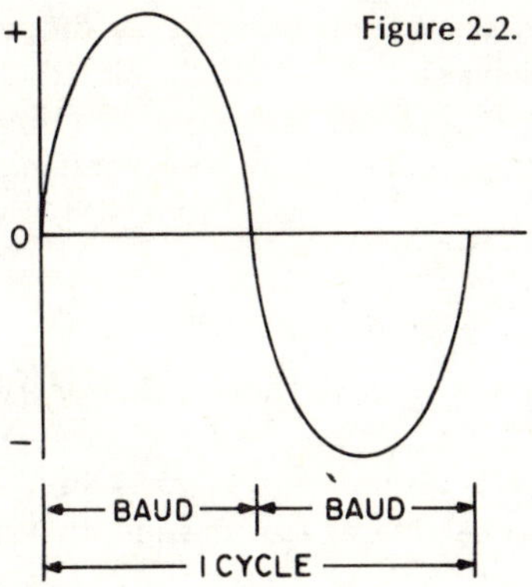


Figure 2-2. *Relationship of bauds and cycles.*

The term "baud" does not give any indication of its information content. As we will learn later, bauds can be treated so that they will carry several bits. If this is done, it is clear that the baud and bit rates will differ.

Another convincing argument against the theory that all bauds carry information bits is the case of the single-frequency tone. Assume a tone of any value — say, 1000 Hz. Since each cycle contains two bauds (see Fig. 2-2), we are "signaling" at a rate of 2000 bauds. But what information can be taken from an uninterrupted single-frequency tone? We might get one bit from the mere presence of the tone itself and maybe another bit or so from its value (1000 Hz), but we must stop there. In summary, then, one may think of bauds as containers traveling along a communication channel, and they may or may not carry bits of information.

### Five-unit Start-Stop Teletypewriter Operations

Table 2-6 gives many interesting relationships between the pulse lengths and operating speeds of various teletypewriters, all of which use the same five-bit baudot code. The average number of words per minute shown in the third column is obtained by dividing the operations per minute by six (number of characters per word). Notice that the next-to-last example gives the same values for bauds and transmission speed (in wpm) as the *Lenkurt* article, which was based on the same machine configuration.

## ENCODING INFORMATION

This section will discuss the form and information content of business-machine signals. This background will help in the study of codes used to enable machines to communicate, which is the subject of Chap. 3.

### Symbols

Up to this point the information content of messages has been discussed in very broad terms. A message was defined as the output of some information source. More specifically, this message, or output, could be a symbol or group of symbols, such as bits, letters of the alphabet, numbers, or special-purpose characters. In addition to the characters themselves, symbols can also be the timing or the spacing between characters, or even the space itself.

### Definition of Character

A character, according to the *American College Dictionary*, is "a symbol as used in a writing system. . . ." It could be a letter, a number, or a symbol with a special meaning, such as a period or a space. When characters are coded for adaptation to data processing or transmission, they are assigned a discrete combination of bits.

**TABLE 2-6. Five-unit Start-stop Telegraph Codes**

| CODE PATTERN (TOTAL NUMBER OF PULSES PER CHARACTER) | NOMINAL SPEEDS AND PULSE LENGTHS: TRANSMITTING SPEEDS | | | PULSE LENGTHS IN MILLISECONDS | | | | WHERE USED |
|---|---|---|---|---|---|---|---|---|
| | OPER. PER MINUTE | AVG. WORDS PER MINUTE | BAUDS | START AND FIVE CODE PULSES | REST PULSE | REC. SHAFT SPEED IN RPM | M SEC. PER CHAR-ACTER | |
| 7.42-Unit | 368* | 61.33 | 45.45 | 22 | 31 | 420 | 163 | Bell System – U.S. |
| 7-Unit | 390* | 65 | 45.45 | 22 | 22 | 420 | 154 | Western Union – U.S. |
| 7.5-Unit | 400** | 66.67 | 50 | 20 | 30 | 461.5 | 150 | C.C.I.T. Standard – Europe |
| 7.42-Unit | 404** | 67.33 | 50 | 20 | 28.4 | 461.7 | 148.4 | U.S. Military, for Inter-operation with Allies |
| 7-Unit | 428.6** | 71.43 | 50 | 20 | 20 | 461.5 | 140 | Former C.C.I.T. Standard – Europe |
| 7.42-Unit | 460 | 76.67 | 56.88 | 17.57 | 25 | 525.7 | 130.43 | U.S. – All Commercial and Military Users |
| 7.42-Unit | 600† | 100 | 74.2 | 13.47 | 19.18 | 685 | 100 | U.S. – All Users |
| 7-Unit | 636† | 106 | 74.2 | 13.47 | 13.47 | 685 | 94.3 | U.S. Military – Limited Use |

*These two codes are compatible
**These three codes are compatible
†These two codes are compatible

In an article appearing in *The Western Union Technical Review,* Fred W. Smith gave the following account concerning the odd length (1.42 times an information bit) of the stop, or rest, pulse: "It was found that the operating margin of some of the early printing telegraph apparatus could be improved by making the stop pulse longer than the start and code pulses. Consequently, the rest pulse length was increased to 1.42 times the length of a code pulse. The question is often raised as to how an odd figure such as 1.42 came to be chosen. One plausible-sounding story is that the engineer who designed the first faceplate distributor intended to make a 7.5-unit faceplate. He calculated the angular length of the code and start segments, then turned over the job of laying out the faceplate to a draftsman, with instruction to provide slots between adjacent segments. The draftsman laid out the faceplate as instructed without shortening the angular length of the code and start segments to compensate for the slots. The rest segment was then shortened to compensate for all of the slots and it turned out to be 1.42 times the length of the other six segments. The first models of the faceplate were made before the error was discovered and it was decided that it would be simpler to adjust the motor governor to obtain the speed necessary to give the correct length code pulses than to make a new faceplate. If this story is true, the error proved to be a convenient one when 1800-rpm synchronous motors came into wide use."

A character may be given varying numbers of bits in different coding systems (such as five for baudot and eight for ASCII [1]), but within any single coding system all characters will have the same number of bits.

## Definition of Word

A word, from a computer's point of view, is a group of characters treated as a unit. The word may have a fixed or a variable number of characters, depending on the particular use. The word may consist of letters, numbers, special characters, or a combination of these. As an example, the IBM 1401 computer uses words of any needed length. The programmer inserts a special mark (a "word mark") to indicate the beginning of each group of characters he wishes the computer to treat as a unit.

In telegraphy, on the other hand, the standard word consists of five characters and a space. To give a meaningful value to the term words per minute (wpm), it has become conventional in communications to assign words this equivalent telegraphic value of six characters.

## Definition of Byte

The term "byte" is used to describe a group of consecutive bits that are treated as a unit. Several computers are designed to use byte-sized characters of six or eight bits. The RCA Spectra 70 Series and the IBM 360 Series, for example, use an eight-bit byte which may represent one alphameric character, two decimal digits, or eight binary digits. For further clarification, the Spectra 70 Series associates bytes with other terms:

1 byte = character
2 bytes = half word
4 bytes = word
8 bytes = double word

## Coding

The amount of information that a symbol will carry depends on the number of symbols in the set from which it was chosen. This was illustrated above when three bits were required to define one out of eight symbols, whereas a minimum of five were needed to identify one letter of the alphabet. This conversion of one set of symbols to another is called "coding." If a complex symbol is converted to a group of more simple symbols, or if a group of symbols is converted to a larger group, coding is *upwards.* An example is the conversion of a letter of the alphabet to its equivalent five bits of the baudot code. Coding downward results in fewer symbols, but each contains more information.

[1] ASCII: American Standard Code for Information Interchange [developed by American National Standards Institute (ANSI), formerly known as United States of America Standards Institute (USASI) which was formerly the American Standards Association (ASA)].

**TABLE 2-7. START-STOP TELEGRAPH AND DATA CODES**

| Unit code (total pulses per character) | Code level (intelligence pulses per character) | Speeds and pulse lengths: Transmitting speeds: Characters per second | Characters per minute | Words per minute | Pulse lengths, in milliseconds: Start & code pulses | Stop, or Rest pulse | Maximum modulation rate in bauds | Where used |
|---|---|---|---|---|---|---|---|---|
| 7-unit | 5 | 6.5 | 390 | 65 | 22 | 22 | 45.45 | Western Union |
| 7-unit | 5 | 10.6 | 636 | 106 | 13.47 | 13.47 | 74.2 | U.S. Military |
| 7.42-unit | 5 | 6.13 | 368 | 61.33 | 22 | 31 | 45.45 | U.S. Gov't & Bell System |
| 7.42-unit | 5 | 6.73 | 404 | 67.33 | 20 | 28.4 | 50 | W.U. Telex (some) |
| 7.42-unit | 5 | 7.67 | 460 | 76.67 | 17.57 | 25 | 56.88 | U.S.A. – All users |
| 7.42-unit | 5 | 10.0 | 600 | 100 | 13.47 | 19.18 | 74.2 | U.S.A. – All users |
| 7.5-unit | 5 | 6.67 | 400 | 66.67 | 20 | 30 | 50 | W.U. Telex (most) |
| 8-unit | 6 | 6.13 | 368 | 61.33 | 20.4 | 20.4 | 49.06 | Teletypesetter |
| 8-unit | 6 | 8.33 | 500 | 83.33 | 15 | 15 | 66.67 | 5-A Ticker |
| 9-unit | 7* | 8.33 | 500 | 83.33 | 13.33 | 13.33 | 75 | IBM Perforated Tape & Transmission code (IBM 1050) |
| 9-unit | 7* | 14.8 | 888 | 148 | 7.5 | 7.5 | 133.33 | IBM Perforated Tape & Transmission code (IBM 1050) |
| 9-unit | 6 | 15 | 900 | 150 | 7.41 | 14.81 | 135 | "900" Ticker (NYSE)** |
| 9-unit | 7* | 66.67 | 4004 | 666.67 | .00167 | .00167 | 600 | IBM System 1030 |
| 10-unit | 8* | 10 | 600 | 100 | 10 | 10 | 100 | Model 37 (optional) |
| 10-unit | 8* | 15 | 900 | 150 | 6.67 | 6.67 | 165 | Model 37 (optional) |
| 11-unit | 8* | 6.67 | 400 | 66.67 | 13.64 | 27.72 | 73.33 | Models 33 & 35 (ACII)† |
| 11-unit | 8* | 10 | 600 | 100 | 9.09 | 18.18 | 110 | Models 33 & 35 (ASCII)† |
| 11-unit | 8* | 15 | 900 | 150 | 6.06 | 12.12 | 165 | Model 37 – W.U. (ASCII)† |

Source: "Bits, Bauds, & Modulation Rates," Western Union Bulletin, No. G-1003/367.
*Includes one parity pulse, or bit.
**NYSE: New York Stock Exchange.
†ASCII: American Standard Code for Information Interchange.

**Reversible Coding**

A coding technique, either upward or downward, is regarded as reversible if each message (character, symbol, code group, etc.) is encoded in a way that is unlike any other encoded message in the same system. The letter *V* in Morse Code is represented by three dots and a dash, for example. If the letter *Y* was represented by the same combination of three dots and a dash, the coding scheme would not be reversible, since the decoder would have an arbitrary choice of *Y* and *V*. So that all messages can be decoded to exactly the same form they had before transmission, it is mandatory that the coding scheme used be completely reversible.

**Conversions between Bits, Characters, and Words**

The transmission of both bits and characters are usually measured in terms of so many per second, whereas words are measured in so many per minute. Remember that the number of bits per character, the number of characters per word, and the number of control bits or characters assigned by different systems can vary over a wide range; thus, conversions from one value to another can be confusing.

Conversion factors can simplify the job of translation. The formulas below may be used for any code on any machine. Note that only the information bits of the particular code are used. In the formulas below, bpx = bits per second, cps = characters per second, and wpm = words per minute. (See Table 2-7.)

$$\text{bps} = \frac{\text{wpm} \times (\text{bits/char})}{10} \quad \text{or} \quad \text{cps} \times (\text{bits/char})$$

$$\text{wpm} = \frac{\text{bps} \times 10}{\text{bits/char}} \quad \text{or} \quad \text{cps} \times 10$$

$$\text{cps} = \frac{\text{bps}}{\text{bits/char}} \quad \text{or} \quad \frac{\text{wpm}}{10}$$

$$\text{bauds} = \frac{1}{\text{length of info pulse (ms)}}$$

**Problems**

1. What are the first six place values of the octal numbering system? Show derivation.
2. Convert the octal number 5372 to its decimal equivalent. Show work.
3. Write, in subscript notation, binary eleven, octal twenty-five, decimal forty-two, and hexadecimal ninety-six.
4. What is the largest digit in the numbering system with a base of 7?
5. Convert $57_{10}$ to its binary equivalent. Show work.
6. Check the answer to Problem 5 by converting it back to decimal. Show work.
7. Add $1100101_2$ and $1100110_2$.

8. A code has eight bits per character. How many different characters can be represented by this code?
9. If one of the eight bits in the code described in Problem 8 was used for some other purpose (such as error detection), how many characters could be described by the remaining seven bits?
10. A model 35 Teletypewriter using the ASCII code and operating at 100 words per minute has a baud rate of 110. What is the length, in milliseconds, of each bit pulse? (1/X equals 110, 1/110 equals X, X equals 9.09 ms.)

*CHAPTER 3*

# *Coding for Communications*

## THE ROLE OF THE BIT

### Background

Early experimenters and mathematicians in the computer field tried many natural and artificial numbering systems in their search for the ideal computer language. Although the binary system was (and is) extremely clumsy to work with manually, the speed of the electronic computer compensated for this disadvantage by performing millions of binary operations per second. This, plus the fact that most electronic devices are inherently two-state (on-off, plus-minus, etc.) and so easily lend themselves to binary operation, led to binary as the basic computing and communicating numbering system.

### The Smallest Unit of Information

A study of information theory will reveal that a bit (either the 1 or the 0) is the smallest unit of information, much in the same way that an atom is the smallest particle of matter. In paper tape or punched cards, the value of a bit is represented by the presence or absence of punched holes; a hole is a 1 bit, and a no-hole is a 0 bit. In magnetic media, such as magnetic tape or magnetic disks, the 1 and 0 bits are represented by magnetized or demagnetized spots, respectively. In the core memory of a computer, the value of a bit is indicated by the direction of the magnetic field in a doughnut-shaped ring.

Bits are combined in various patterns to create characters, and this process is called "coding." The following paragraphs will explain many of the codes in common use. The text is divided into sections, each of which discusses a separate grouping of codes, such as those for paper tape, those for punched cards, and those used with magnetic tape and computers. Examples of various kinds of code construction are detailed in the text, and are listed in the code tables at the end of the chapter.

Although *any* binary code can be transmitted, two or three codes have become popular to the point that all others are nearly totally excluded. These codes owe their popularity to their relatively high efficiency (high average information content per bit transmitted). A certain code also may be used to the exclusion of all others because it is an inherent or congenital part of the equipment. Model 28 teletypewriters, for example, use a five-bit code with characters and functions specifically and mutually designed into the system. Any unilateral changes to the code or the machines would result in errors or malfunctions in the system.

Some transmission systems are "code transparent": they are insensitive to the code structure and will transmit and receive any code used. It follows, however, that such transmission systems cannot generate or interpret these codes, but merely move them from one point to another. The generation and later interpretation or use of the data transmitted through a transparent system must be left to machines outside the transmission system. The Bell System's Dataspeed units, which transmit and receive any code up to eight bits per character, are examples of transparent equipment.

## PAPER-TAPE CODING

### Introduction

Punched paper tapes are generally conceded to be the least expensive medium for storing data in machine language. These tapes are widely used in applications which call for relatively short storage times, limited processing of the stored data, and relatively infrequent use. Punched paper tape is used most commonly with teletypewriter systems. Messages are "cut" into paper tape manually, then sent at machine speeds to reduce circuit time. Incoming messages are sometimes cut into tape simultaneously as the message is printed, and the tapes are used to forward the message by way of another teletypewriter circuit or may be used as input to another machine, such as a computer.

Paper tapes are also widely used in office machines that perform repetitive work. The text of a letter, for example, can be cut into paper tape on a machine resembling a typewriter. When the same tape is run back through, or "read" by the machine, it will produce the same error-free typed letter, and the operator merely inserts the individual headings.

There are many more applications for punched paper tape and the other data storage media, but this chapter deals with the codes themselves. The following paragraphs briefly describe five-channel and eight-channel paper-tape codes and parity in general; then several common codes are discussed in detail.

As mentioned above, the presence of a hole in punched paper tape represents a binary 1, or a *mark,* whereas a space not punched indicates a binary 0, or space. Marks and spaces are combined in various ways to construct codes.

**Five-channel Tape**

The tape shown in Fig. 3-1 contains characters in a five-channel code. The channels (or levels) run the *length* of the tape, whereas each coded character occupies a column across the *width* of the tape. The feed-holes, which are engaged by the drive-wheel of the tape-handling mechanism, run

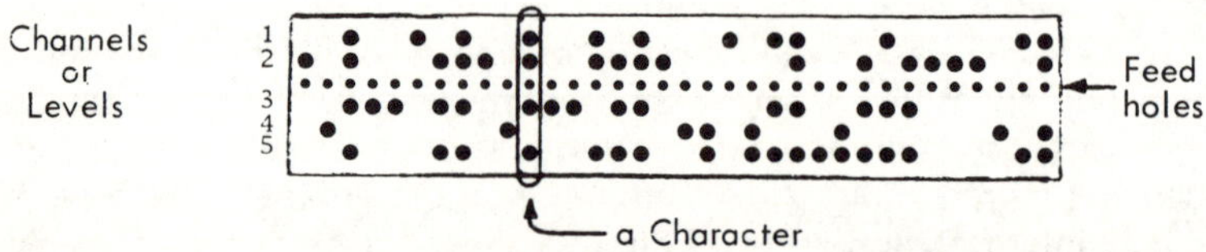


Figure 3-1. *Five-channel paper tape.*

between the second and third channels. A five-channel code requires all five bits to create a character, leaving no way to introduce an extra bit in each character to check for errors. To overcome this weakness and to expand the number of characters which can be represented, codes of six, seven, and eight channels were developed.

**Eight-channel Tape**

An eight-channel paper tape is shown in Fig. 3-2. As with the five-channel tape, the channels run the length of the tape, and the characters are punched across it. The feed-holes in an eight-channel tape are between the third and fourth channels. Information bits occupy the first six or seven channels, depending on the coding. The seventh (or eighth) channel is used for a "parity" bit, which is used for error detection.

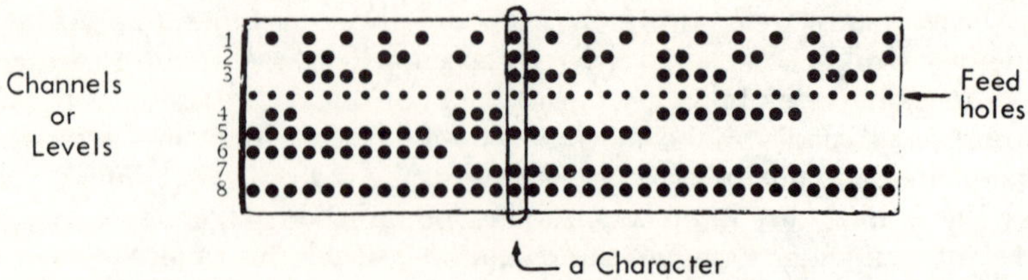


Figure 3-2. *Eight-channel paper tape.*

### Errors and Parity

In the transmission of data, an error has occurred if the sequence of received bits is not the same as the transmitted sequence. Such a condition is the result of bits being lost, changed in value, or rearranged. The factors which cause such disturbances will be discussed in the following chapter, but the point here is that they cannot be prevented. Thus, there is always the chance that errors will occur during the transmission of data.

Since accuracy is of prime importance in data processing, some method is required to determine if data has maintained its desired value or form during processing, transmission, or storage. In paper tape codes, a parity bit is added to the information bits as an error-checking device. A parity bit is "a check bit that indicates whether the total number of binary '1' digits in a character or word (excluding the parity bit) is odd or even. If a '1' parity bit indicates an odd number of '1' digits, then a '0' bit indicates an even number of them. If the total number of '1' bits, including the parity bit, is always even, the system is called an even parity system. In an odd parity system, the total number of '1' bits, including the parity bit, is always odd." [1]

### Baudot Code

The baudot code (Fig. 3-3) is a five-channel code consisting of an arbitrary arrangement of marks and spaces. This lack of sequential relationship between characters makes the code "noncomputable." In addition, observation of the arrangement of the marks and spaces of any single character is not sufficient to permit recognition of its meaning, since the last preceding shift code must be known to determine whether the character is upper or lower case.

There are 32 different possible combinations of marks and spaces in a five-channel tape ($2^5 = 32$). With the use of an upper case, the number of usable combinations increases to 58 ($32 \times 2$, less 6 function codes: space, figs, carriage return, ltrs, blank, and line feed). Since the alphabet, the numbers, and the functions require 42 of the 58 combinations, the baudot code is severely limited in extra punctuation, function, and computing codes.

For transmission of intelligence to distant points, teletypewriters convert the baudot code into direct-current (d-c) pulses. The flow of current on the line represents a mark, and a space is represented by a no-current condition, as shown in Fig. 3-4.

Figure 3-5 shows the configuration of an entire character as it would appear on the line. The intelligence portion of the character is its baudot coding (in this case, *Y*). Each mark and space is an equal time interval, so that the combinations of marks and spaces making up any character require the same length of time to transmit. The start and stop intervals, which provide

[1] *Automatic Data Processing Glossary*, Superintendent of Documents. Washington, D. C.: U. S. Gov't Printing Office, 1962.

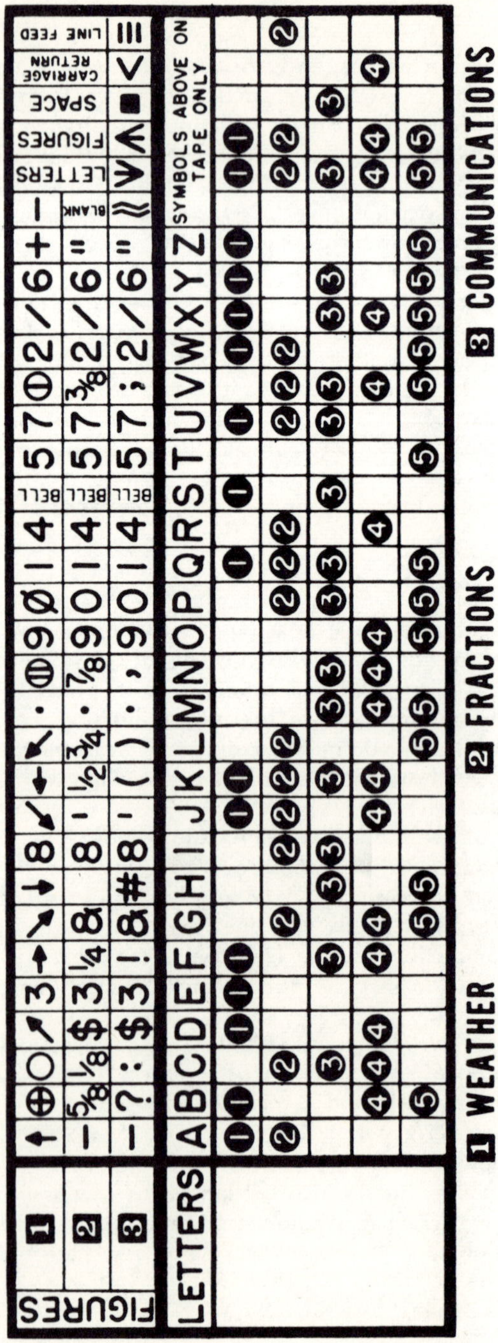


Figure 3-3. *Five-channel baudot telegraphy code.*

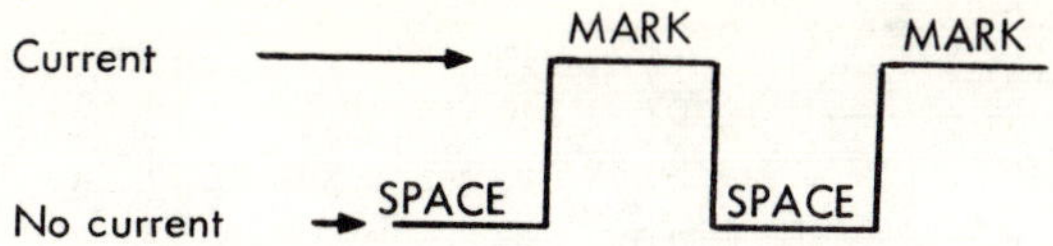


Figure 3-4. *Mark and space current conditions.*

synchronization between machines, are added to the baudot coding as part of the line signal of every character. The mark and space intervals may be assigned an arbitrary time value of 1 for purposes of illustration. The start interval thus also has a time value of 1 (from Fig. 3-5), but the stop interval is longer, with a value of 1.42 times the length of the others. Any character in the five-channel telegraphic code, therefore, consists of 7.42 time intervals: one start, five intelligence, and 1.42 stop. Although all 7.42 segments of the baudot code are commonly called bits, only the five that represent the coded character carry information — the start and stop segments do not carry information. (See Chap. 2.)

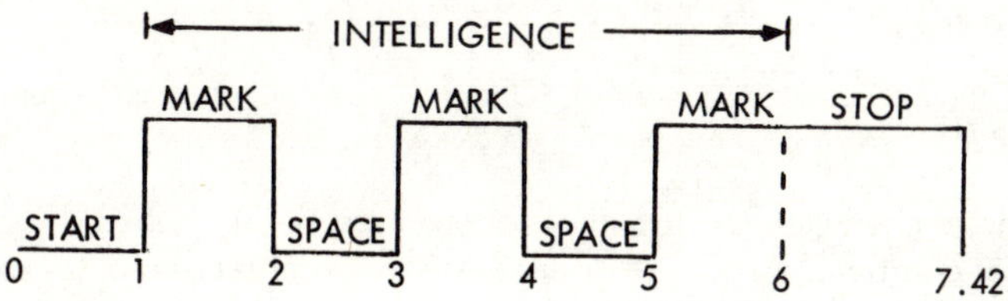


Figure 3-5. *Line configuration of a five-bit baudot character (Y).*

**ASCII Code**

Because of the limited number of possible code combinations of the baudot code and, more important, because of its lack of a logical or sequential pattern, new codes which are more flexible and which readily lend themselves to computing have been developed. Most of these are six- or seven-channel codes, which use an additional channel for parity.

One of these newer codes that are rapidly gaining acceptance is the ASCII code — American Standard Code for Information Interchange. It is also called the ANSI code and the Data Interchange code. It is a seven-channel code which adds an eighth channel to provide for an even parity.

The bit configuration of the ASCII code is shown in Fig. 3-6. The graphics (printable characters) and control characters (functions) are listed by columns within the double lines of the chart. At the top of each column is the high-order bit configuration for any character or function in that column. To the left of each row is given the configuration of the four low-order bits, which describe each of the characters or functions in that row.

| Bits b7 → b6 → b5 → | | | | | 0 0 0 | 0 0 1 | 0 1 0 | 0 1 1 | 1 0 0 | 1 0 1 | 1 1 0 | 1 1 1 |
|---|---|---|---|---|---|---|---|---|---|---|---|---|
| b4 ↓ | b3 ↓ | b2 ↓ | b1 ↓ | Column → Row ↓ | 0 | 1 | 2 | 3 | 4 | 5 | 6 | 7 |
| 0 | 0 | 0 | 0 | 0 | NUL | DLE | SP | 0 | @ | P | ` | p |
| 0 | 0 | 0 | 1 | 1 | SOH | DC1 | ! | 1 | A | Q | a | q |
| 0 | 0 | 1 | 0 | 2 | STX | DC2 | " | 2 | B | R | b | r |
| 0 | 0 | 1 | 1 | 3 | ETX | DC3 | # | 3 | C | S | c | s |
| 0 | 1 | 0 | 0 | 4 | EOT | DC4 | $ | 4 | D | T | d | t |
| 0 | 1 | 0 | 1 | 5 | ENQ | NAK | % | 5 | E | U | e | u |
| 0 | 1 | 1 | 0 | 6 | ACK | SYN | & | 6 | F | V | f | v |
| 0 | 1 | 1 | 1 | 7 | BEL | ETB | ' | 7 | G | W | g | w |
| 1 | 0 | 0 | 0 | 8 | BS | CAN | ( | 8 | H | X | h | x |
| 1 | 0 | 0 | 1 | 9 | HT | EM | ) | 9 | I | Y | i | y |
| 1 | 0 | 1 | 0 | 10 | LF | SUB | * | : | J | Z | j | z |
| 1 | 0 | 1 | 1 | 11 | VT | ESC | + | ; | K | [ | k | { |
| 1 | 1 | 0 | 0 | 12 | FF | FS | , | < | L | \ | l | \| |
| 1 | 1 | 0 | 1 | 13 | CR | GS | — | = | M | ] | m | } |
| 1 | 1 | 1 | 0 | 14 | SO | RS | . | > | N | ^ | n | ~ |
| 1 | 1 | 1 | 1 | 15 | SI | US | / | ? | O | — | o | DEL |

Figure 3-6. *Chart of American Standard Code for Information Interchange (ASCII).*

Any single character or function, then, can be described by its own particular combination of high-order and low-order bits. Find, for example, the code for the letter *F*. (See Fig. 3-7.)

Notice how both the high-order and the low-order bit groups increase in value by straight binary addition and how the graphics and controls are grouped according to function. In contrast to the arbitrary baudot code, these logical relationships of the ASCII code classify it as a "computable" code. Since the letters of the alphabet, plus all the other characters, are assigned values in a continuous binary sequence, the computer does not have to leave its native binary language to perform sequential operations with these characters. And, because of the grouping of the control functions, the computer must analyze only the first two bits of a code to determine whether it is a control (both zeros) or a character (not both zeros).

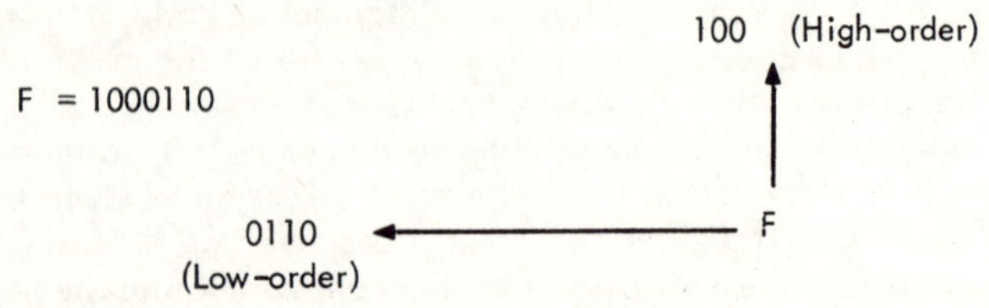


Figure 3-7. *Interpreting the ASCII code chart.*

## ASCII CONTROL CHARACTER DEFINITIONS[2]

**NUL (Null):** The all-zeros character which may serve to accomplish time fill and media fill.

**SOH (Start of Heading):** A communication control character used at the beginning of a sequence of characters which constitute a machine-sensible address or routing information. Such a sequence is referred to as the heading. An STX character has the effect of terminating a heading.

**STX (Start of Text):** A communication control character which precedes a sequence of characters that is to be treated as an entity and entirely transmitted through to the ultimate destination. Such a sequence is referred to as "text." STX may be used to terminate a sequence of characters started by SOH.

**ETX (End of Text):** A communication control character used to terminate a sequence of characters started with STX and transmitted as an entity.

**EOT (End of Transmission):** A communication control character used to indicate the conclusion of a transmission, which may have contained one or more texts and any associated headings.

**ENQ (Enquiry):** A communication control character used in data communication systems as a request for a response from a remote station. It may be used as a "Who Are You" (WRU) to obtain identification or to obtain station status, or both.

**ACK (Acknowledge):** A communication control character transmitted by a receiver as an affirmative response to a sender.

**BEL (Bell):** A character for use when there is a need to call for human attention. It may control alarm or attention devices.

**BS (Backspace):** A format effector which controls the movement of the printing position one printing space backward on the same printing line. (Applicable also to display devices.)

**HT (Horizontal Tabulation):** A format effector which controls the movement of the printing position to the next in a series of predetermined positions along the printing line. (Applicable also to display devices and the skip function on punched cards.)

**LF (Line Feed):** A format effector which controls the movement of the printing position to the next printing line. (Applicable also to display devices.)

**VT (Vertical Tabulation):** A format effector which controls the

[2] The above definitions are recommendations by Sectional Committee X3 of the American National Standards Institute (ANSI). This standard does not define the means by which the coded set is to be recorded in any physical medium, nor does it include any redundancy or define techniques for error control. Further, this standard does not define data communication character structure, data communication formats, code extension techniques, or graphic representation of control characters. And, since revisions are still being suggested for this code, several versions are in existence. Care should be taken to reach mutual agreement on the version to be used in any particular application.

movement of the printing position to the next in a series of predetermined printing lines. (Applicable also to display devices.)

**FF (Form Feed):** A format effector which controls the movement of the printing position to the first predetermined printing line on the next form or page. (Applicable also to display devices.)

**CR (Carriage Return):** A format effector which controls the movement of the printing position to the first printing position on the same printing line. (Applicable also to display devices.)

**SO (Shift Out):** A control character indicating that the code combinations which follow shall be interpreted as outside of the character set of the standard code table until a *Shift In* character is reached.

**SI (Shift In):** A control character indicating that the code combinations which follow shall be interpreted according to the standard code table.

**DLE (Data Link Escape):** A communication control character which will change the meaning of a limited number of contiguously following characters. It is used exclusively to provide supplementary controls in data communication networks.

**DC1, DC2, DC3, DC4 (Device Controls):** Characters for the control of ancillary devices associated with data processing or telecommunication systems, more especially switching devices "on" or "off." (If a single "stop" control is required to interrupt or turn off ancillary devices, DC4 is he preferred assignment.)

**NAK (Negative Acknowledge):** A communication control character transmitted by a receiver as a negative response to the sender.

**SYN (Synchronous Idle):** A communication control character used by a synchronous transmission system in the absence of any other character to provide a signal from which synchronism may be achieved or retained.

**ETB (End of Transmission Block):** A communication control character used to indicate the end of a block of data for communication purposes. ETB is used for blocking data where the block structure is not necessarily related to the processing format.

**CAN (Cancel):** A control character used to indicate that the data with which it is sent is in error or is to be disregarded.

**EM (End of Medium):** A control character associated with the sent data which may be used to identify the physical end of the medium or the end of the used, or wanted, portion of information recorded on a medium. (The position of this character does not necessarily correspond to the physical end of the medium.)

**SUB (Substitute):** A character that may be substituted for a character which is determined to be invalid or in error.

**ESC (Escape):** A control character intended to provide code extension (supplementary characters) in general information interchange.

The escape character itself is a prefix affecting the interpretation of a limited number of contiguously following characters.

**FS (File Separator), GS (Group Separator), RS (Record Separator), and US (Unit Separator):** These information separators may be used within data in optional fashion, except that their hierarchical relationship shall be: FS is the most inclusive, then GS, then RS, and US is least inclusive. (The content and length of a file, group, record, or unit are not specified.)

**DEL (Delete):** This character is used primarily to erase or obliterate erroneous or unwanted characters in perforated tape. (In the strict sense, DEL is not a control character.)

**SP (Space):** A normally nonprinting graphic character used to separate words. It is also a format effector which controls the movement of the printing position, one printing position forward. (Applicable also to display devices.)

The paper tape form of one version of the ASCII code is shown in Fig. 3-8. The binary sequence (such as the 1000110, which was the letter F in an example above) has been converted to a series of punched holes. Note, too, the use of the eighth or bottom channel for even parity. The abbreviations for some of the control functions do not resemble those in the above example, although their functions are nearly all the same, because this example is from an earlier version of the code. Changes will continue to be made until a standard has been accepted. Even then, users are free to give any meaning to any particular character configuration when the code will be used only within the user's own system.

## Other Paper Tape Codes

Several other paper tape codes are used today, but less widely than baudot or ASCII. An international code approved by the International Telephone and Telegraph Consultative Committee (CCITT) is the CCITT #2. It closely resembles baudot, differing only in some of the special characters. IBM has named its transmission code EBCDIC: Extended Binary Coded Decimal Interchange Code. EBCDIC is constructed in binary sequence as is ASCII, but it lacks a parity bit because all eight bits are used for information. Its collating sequence of letters before numbers is the reverse of that used with ASCII.

Two almost identical codes are Friden's Flexowriter code and the one developed for use on the Mach-10 communication typewriter. They differ on a few special characters. The Field Data code, developed by the military as a compromise between codes of several machine vendors, incorporates the good features of each. Its construction in binary sequence lends it to computing, and thus overcomes the most serious disadvantage of baudot. Its

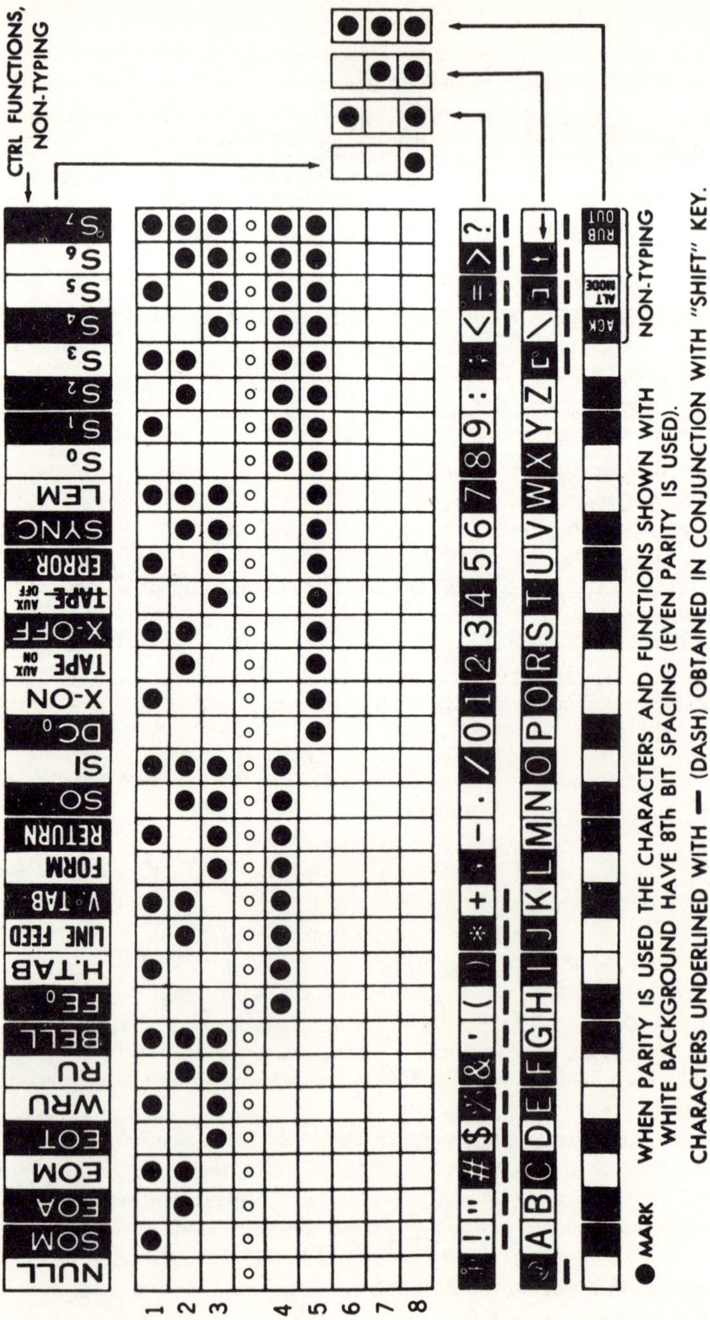


Figure 3-8. *A paper tape for the one version of ASCII.*

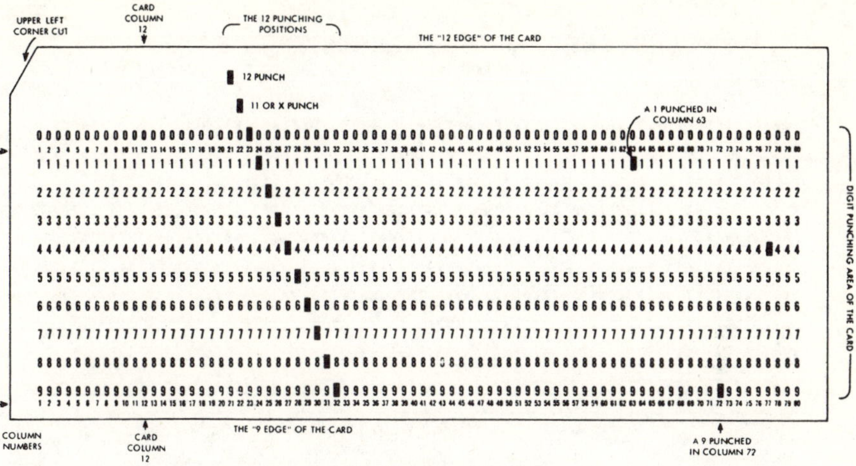


Figure 3-9. *An IBM punched card.*

characters consist of only six bits, which makes it more economical than EBCDIC or ASCII, but it lacks a parity bit.

The structure of each of the codes discussed above is described in Table 3-3 at the end of this chapter.

## PUNCHED-CARD CODES

As with paper tape, punched cards (often called "unit records") carry information by having holes punched in them in certain patterns. (See Fig. 3-9.) But punched cards do not have the wide variety of codes that paper tape does. In fact, with the exception of "edge-punched" cards which will be discussed later, two codes dominate the punched-card area.

### Hollerith Code

The most widely used code in punched cards is the hollerith code, which is shown in Fig. 3-10 on an 80-column card. Because hollerith is arranged in a logical pattern, it lends itself very well to sorting, filing, computing, and the like.

The card format consists of 80 vertical columns, each of which can carry one digit, letter, or special character. The columns can be divided into several groups of varying lengths, called fields. Each field can represent a particular piece of information, such as a customer's name, his identification number, an invoice number, a quantity, a dollar amount, and so on. Usually one punched card represents a record of one transaction, thus the cards are often referred to as "unit records."

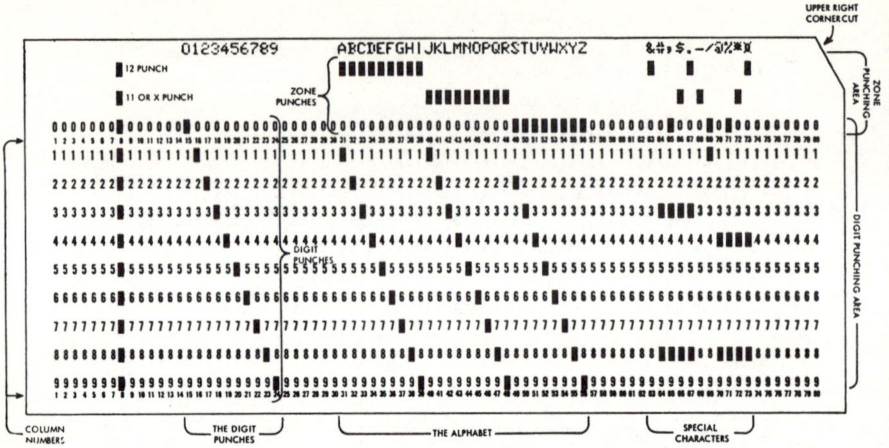


Figure 3-10. *Hollerith punched-card code.*

## Remington Rand Code

Another popular punched-card code is that used by Remington Rand (Fig. 3-11). Its most obvious characteristic is its arrangement in two horizontal rows along the length of a card. Each row has 45 columns, giving the card a capacity of 90 columns, or characters. A card carrying the Remington Rand code is usually punched with round holes, rather than the rectangular holes of the hollerith card. An application of the Remington Rand card appears in Fig. 3-12.

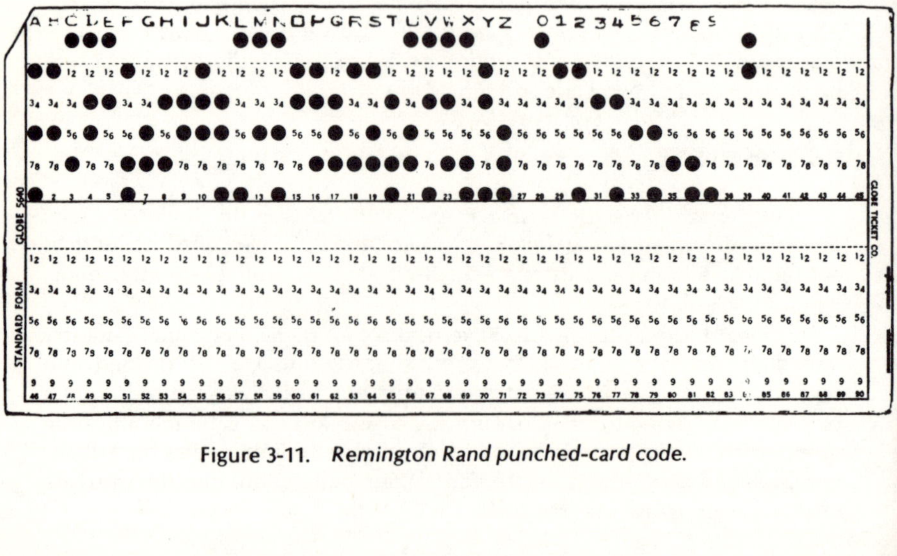


Figure 3-11. *Remington Rand punched-card code.*

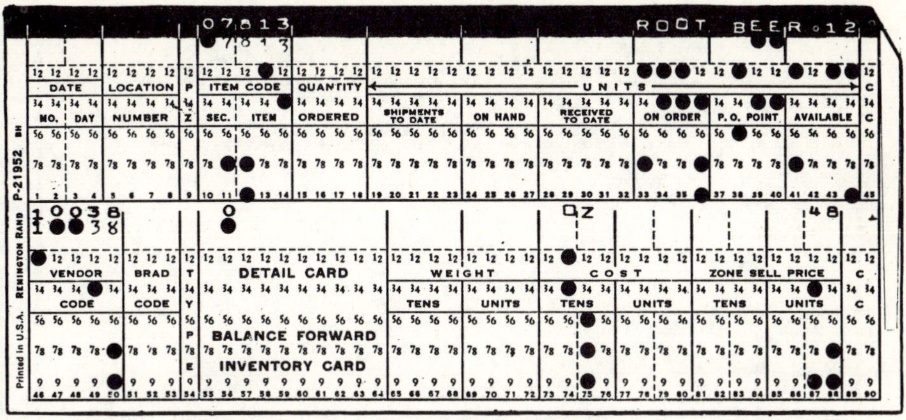


Figure 3-12. *An application of the Remington Rand card.*

### Edge-punched Cards

Edge-punched cards (see Figs. 3-13 and 3-14) are used by some systems to combine the advantages of paper tape and the unit-record punched card. The cards, approximately the same size as a standard punched card, have a paper-tape code punched along one edge, and are sometimes connected at the ends to form a fan-fold. Several business machines which can read paper tape, such as Friden's Flexowriter, have attachments which permit them to handle edge-punched cards containing the machine's native paper-tape language.

Figure 3-13. *A Friden edge-punched card.*

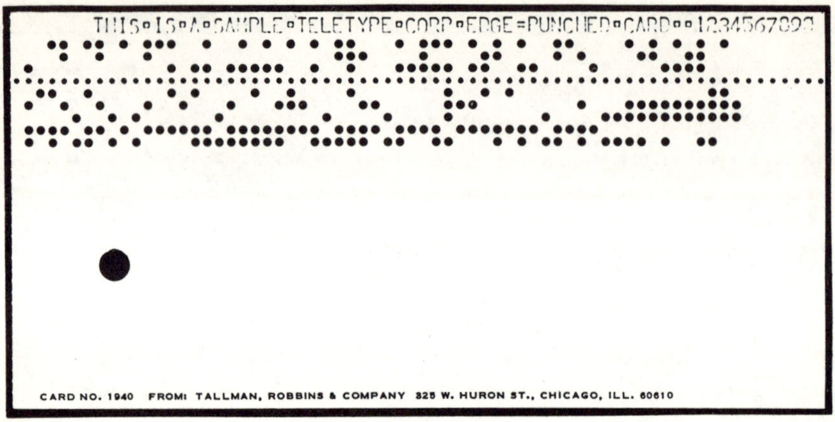


Figure 3-14. *A Teletype Corp. edge-punched card.*

## Packed Decimal

The term "packed decimal" refers to a grouping of two or more decimal digits in binary notation into a representation normally used for only one digit. An eight-bit ASCII digit completely fills an eight-bit byte, for example, while two decimal digits, each represented by four bits, could be packed into one such byte.

The card in Fig. 3-15 is a sample of one of many methods of packing decimal digits. To interpret the card, note the positions of the punched holes. The first column is punched in positions 12, 11, 0, 5, and 9. Refer to Fig. 3-16 and find 9, 12, 11, 0 as zone punches, and 5 as a digit punch, in the first chart. At the opposite ends of this column and row, find 0111 and 0101, the

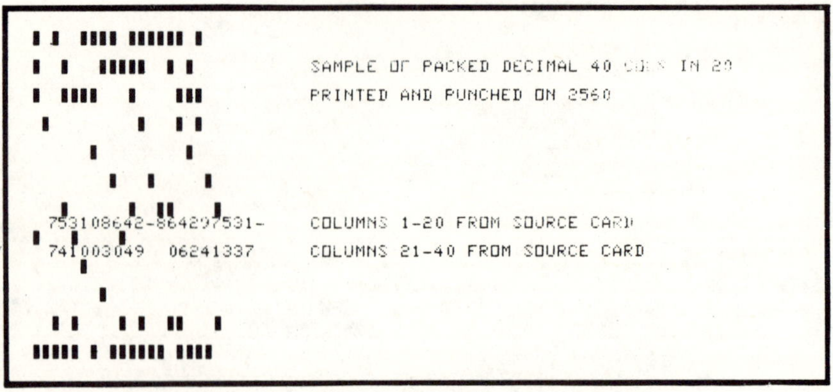


Figure 3-15. *A punched card containing packed decimal.*

binary representation of decimal 7 and 5. The first column of the punched card thus contains the decimal digits 7 and 5.

Decimal digits can be even more packed on a card. Since a punched card has twelve punch positions in each column, and only four bits are required to represent any decimal digit in binary form, it follows that three decimal digits can be packed into one card column. One argument against this method is that it is possible to punch so many holes in a card that it becomes weak and thus hard to handle. The packing of triple decimal digits into one byte is very useful, however, in the transmission of data when the data is being transferred directly from machine to machine; up to two-thirds of the transmission time can be saved.

## MAGNETIC TAPE AND COMPUTER CODING

### Introduction

There are some computer codes not often used for transmission that may be of interest due to their structure. This section discusses a few codes used in computers and their peripheral units.

Computers use magnetic (mag) tape much in the same way as does a home tape recorder — the tapes are used to collect and store the information put on them and will retain that information for repeated usage until changed or erased.

The configuration of characters recorded on magnetic tape closely resemble those on paper tape, but with some significant differences. The rows, called "channels" on paper tape, are "tracks" on magnetic tape, and magnetic tape carries from four to nine tracks. Whereas paper-tape codes space characters ten to the inch, magnetic tape usually carries 200, 556, 800, or 1600 to the inch.

Magnetic tapes are made of a plastic material with a coating of a metallic oxide. The character bits which would appear as holes in a paper tape (the 1 bits) are in the form of magnetized spots in the metallic oxide, whereas the 0 bits are not magnetized. Some high-density magnetic tapes are magnetized for both the 1 and the 0 bits; a flux change indicates which value is recorded (see section on baseband signals). When a character is read from magnetic tape to the memory unit of a computer, it usually retains its bit configuration. The following codes, then, are used for both magnetic tapes and computers and are sometimes suited for communicating between machines.

### Binary Coded Decimal (BCD)

Because pure binary is so cumbersome, especially in the larger numbers, the binary coded decimal (BCD) code was developed for computer use. It is a compromise between the decimal figures humans are accustomed to and the pure binary language of the computer or other business machine.

| BIT POSITIONS 0, 1 → | 00 | | | | 01 | | | | |
|---|---|---|---|---|---|---|---|---|---|
| BIT POSITIONS 2, 3 → | 00 | 01 | 10 | 11 | 00 | 01 | 10 | 11 | DIGIT PUNCHES |
| BIT POSITIONS 4, 5, 6, 7: 0000 | (1) NUL | (2) | (3) DS | (4) | (5) SP | (6) & | (7) – | (8) | |
| 0001 | | | SOS | | | | / (13) | | 1 |
| 0010 | | | FS | | | | | | 2 |
| 0011 | | TM | | | | | | | 3 |
| 0100 | PF | RES | BYP | PN | | | | | 4 |
| 0101 | HT | NL | LF | RS | | | | | 5 |
| 0110 | LC | BS | EOB | UC | | | | | 6 |
| 0111 | DL | IL | PRE | EOT | | | | | 7 |
| 1000 | | | | | | | | | 8 |
| ZONE PUNCHES | 9 12 | 9 11 | 9 0 | 9 | 9 12 0 | 9 12 11 | 9 11 0 | 9 12 11 0 | |

| BIT POSITIONS 0, 1 → | 10 | | | | 11 | | | | |
|---|---|---|---|---|---|---|---|---|---|
| BIT POSITIONS 2, 3 → | 00 | 01 | 10 | 11 | 00 | 01 | 10 | 11 | DIGIT PUNCHES |
| BIT POSITIONS 4, 5, 6, 7: 0000 | | | | | (9) | (10) | (11) | (12) 0 | 8-1 |
| 0001 | a | j | | | A | J | (14) | 1 | 1 |
| 0010 | b | k | s | | B | K | S | 2 | 2 |
| 0011 | c | l | t | | C | L | T | 3 | 3 |
| 0100 | d | m | u | | D | M | U | 4 | 4 |
| 0101 | e | n | v | | E | N | V | 5 | 5 |
| 0110 | f | o | w | | F | O | W | 6 | 6 |
| 0111 | g | p | x | | G | P | X | 7 | 7 |
| 1000 | h | q | y | | H | Q | Y | 8 | 8 |
| 1001 | i | r | z | | I | R | Z | 9 | 9 |
| ZONE PUNCHES | 12 0 | 12 11 | 11 0 | 12 11 0 | 12 | 11 | 0 | | |

| Bit positions 0, 1 → | 00 | | | | 01 | | | | |
|---|---|---|---|---|---|---|---|---|---|
| **Bit positions 2, 3 →** | **00** | **01** | **10** | **11** | **00** | **01** | **10** | **11** | |
| **Bit positions 4, 5, 6, 7** | | | | | | | | | **Digit punches** |
| 1001 | | | | | | | | | 8-1 |
| 1010 | | CC | SM | | ] | [ | ⑮ | : | 8-2 |
| 1011 | | | | | . | $ | , | # | 8-3 |
| 1100 | | | | | < | * | % | @ | 8-4 |
| 1101 | | | | | ( | ) | _ | ' | 8-5 |
| 1110 | | | | | + | ; | > | = | 8-6 |
| 1111 | | | | | ! | ^ | ? | " | 8-7 |
| Zone punches | 9 12 | 9 11 | 9 0 | 9 | 12 | 11 | 0 | | |

| Bit positions 0, 1 → | 10 | | | | 11 | | | | |
|---|---|---|---|---|---|---|---|---|---|
| **Positions 2, 3 →** | **00** | **01** | **10** | **11** | **00** | **01** | **10** | **11** | |
| **Bit positions 4, 5, 6, 7** | | | | | | | | | **Digit punches** |
| 1010 | | | | | | | | | 8-2 |
| 1011 | | | | | | | | | 8-3 |
| 1100 | | | | | | | | | 8-4 |
| 1101 | | | | | | | | | 8-5 |
| 1110 | | | | | | | | | 8-6 |
| 1111 | | | | | | | | | 8-7 |
| Zone punches | 12 0 | 12 11 | 11 0 | 12 11 0 | 9 12 0 | 9 12 11 | 9 11 0 | 9 12 11 0 | |

① 12-0-9-8-1
② 12-11-9-8-1
③ 11-0-9-8-1
④ 12-11-0-9-8-1
⑤ NO PUNCHES
⑥ 12
⑦ 11
⑧ 12-11-0
⑨ 12-0
⑩ 11-0
⑪ 0-8-2
⑫ 0
⑬ 0-1
⑭ 11-0-9-1
⑮ 12-11

Figure 3-16. *Encode-decode tables for packed decimal — EBCDIC. (Courtesy of IBM)*

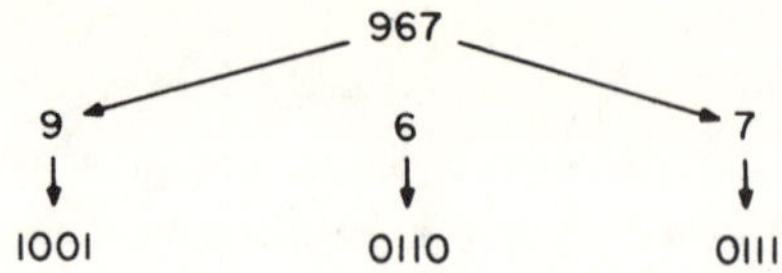


Figure 3-17. *Conversion of decimal to binary-coded decimal — BCD.*

BCD is generated by separating the digits of a number, then giving each digit its own binary equivalent. High-order zeros are added to each binary number if necessary to give it four digits. For example, the number 967 has a pure binary equivalent of 1111000111. The BCD system first separates the three decimal digits of the number, then gives each its own binary equivalent, as shown in Fig. 3-17.

The advantage of BCD over pure binary is that only the binary equivalents of the base number (0 through 9) need to be remembered. Table 3-1 gives the binary and BCD equivalents for decimal numbers 0 through 9, plus conversions. An operator, then, can work in BCD, and the machine easily converts it to pure binary internally. BCD has further importance in that it serves as the basis of several other codes used in computing and communications, several of which are discussed below.

**TABLE 3-1. DECIMAL, BINARY, AND BCD EQUIVALENTS**

| Decimal | Binary | BCD |
|---|---|---|
| 0 | 0 | 1010 |
| 1 | 1 | 0001 |
| 2 | 10 | 0010 |
| 3 | 11 | 0011 |
| 4 | 100 | 0100 |
| 5 | 101 | 0101 |
| 6 | 110 | 0110 |
| 7 | 111 | 0111 |
| 8 | 1000 | 1000 |
| 9 | 1001 | 1001 |
| 10 | 1010 | 0001 1010 |
| 15 | 1111 | 0001 0101 |
| 16 | 10000 | 0001 0110 |
| 967 | 1111000111 | 1001 0110 0111 |

BCD (without parity) is represented on magnetic tape by four tracks, with each track assuming a binary place value of 1, 2, 4, or 8. Here is a BCD digit with normal binary place values:

Place value: 8 4 2 1
BCD digit (1): 0 0 0 1

On magnetic tape, the digits have a slightly different form, as shown in Fig. 3-18.

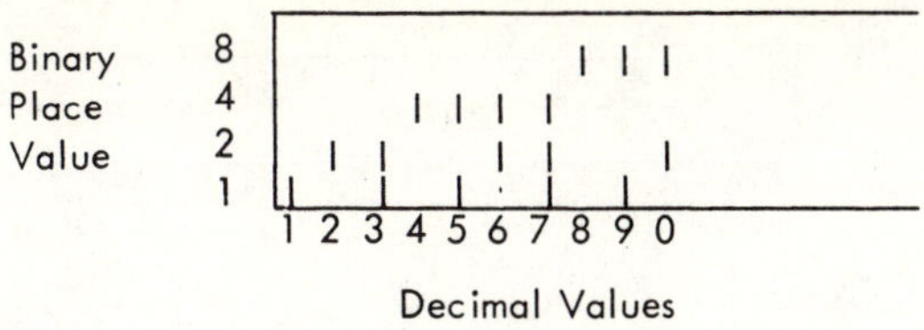


Figure 3-18. *Decimal and binary values on magnetic tape.*

**Excess-3 Code (XS3)**

The Excess-3 computer code shown in Fig. 3-19 is used, among others, by some Univac and RCA machines. The code is based on BCD, but each digit is presented as if it were three larger than its true value: 3 = 6, 5 = 8, and so on. Among the advantages of the Excess-3 code over straight BCD is that the zero has a logical bit configuration of its own.

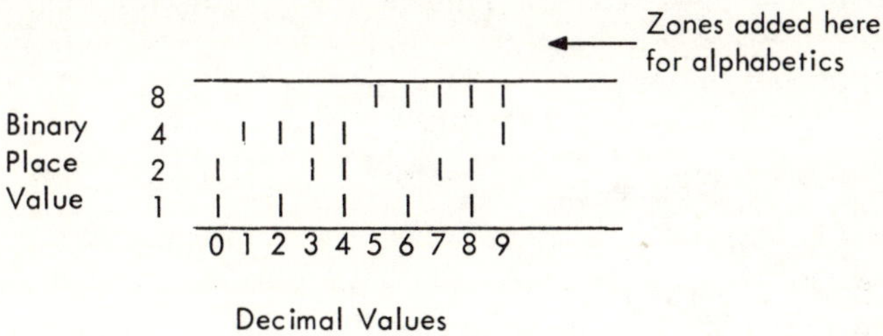


Figure 3-19. *Excess-three code.*

**Binary Coded Octal (BCO)**

Binary coded octal computer code uses the same principles as binary coded decimal, but with the added advantages of having only three positions per digit rather than four (thus making more efficient use of space), and of being easily converted to either pure binary or decimal. (See Table 3-2.)

**TABLE 3-2. BINARY CODED OCTAL**

| Decimal | Octal | BCO | |
|---|---|---|---|
| 0 | 0 | 000 | |
| 1 | 1 | 001 | |
| 2 | 2 | 010 | |
| 3 | 3 | 011 | |
| 4 | 4 | 100 | |
| 5 | 5 | 101 | |
| 6 | 6 | 110 | |
| 7 | 7 | 111 | Binary |
| 8 | 10 | 001 000 | = 1000 |
| 9 | 11 | 001 001 | 1001 |

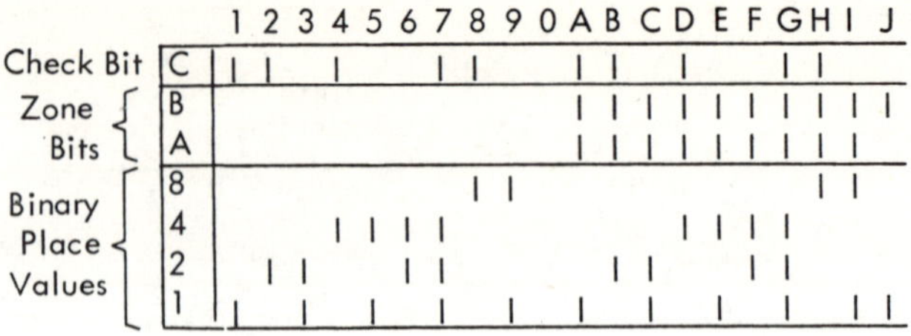


Figure 3-20. *Seven-bit alphanumeric code. Note: (1) The C zone is used for parity. (2) The B and A zones are used for alphabetics. (3) The binary cycle repeats itself from 1 to 9, A to I, J to R, and S to Z. Note, however, that the S is a binary 2, not a 1.*

## Seven-Bit Alphameric Code

A code widely used for magnetic tape, especially by IBM, the seven-bit alphameric code uses BCD for numerics, and adds two zone bits for alphabetics. (See Fig. 3-20.) Another bit is used for even-bit parity, making a total of seven bits per character.

## Two-out-of-five Code

The two-out-of-five computer code deviates from the popular binary sequence, but with a purpose. Its name comes from the way each character has five bits, two of which are always one-bits. The other three are always zeros. The bit pattern is arranged in a logical pattern, which makes the code "computable." But the greatest advantage of the two-out-of-five code over straight BCD or binary is its built-in even parity system of two one-bits per character. The configuration of the zero is arbitrary. (See Fig. 3-21.)

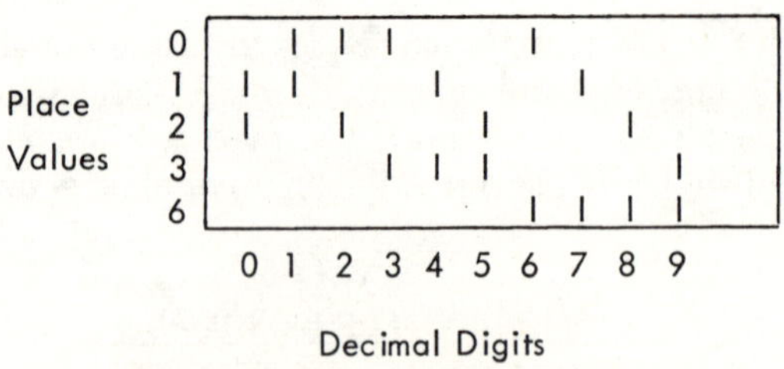


Figure 3-21. *The two-out-of five code.*

## Biquinary Code

Another computer code with built-in even parity is the biquinary code, which is used by the IBM 650 computer. Each character has a seven-bit coding. Of the seven bits, two are always one-bits, and the other five are zero-bits. (See Fig. 3-22.)

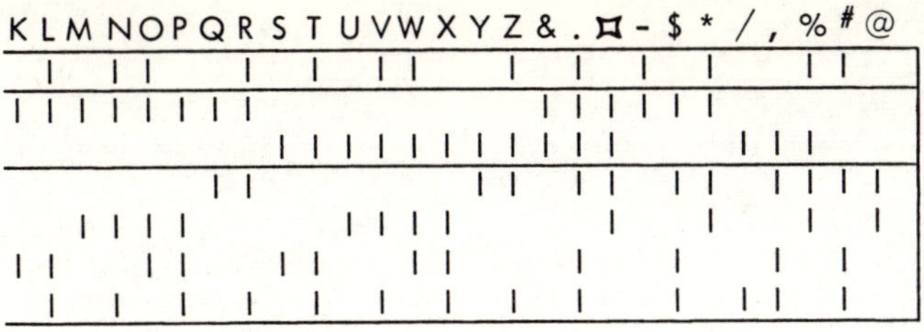


## Table of Common Codes

Table 3-3 contains most of the communications codes in use today. Many of them have been discussed in the preceding pages. Those that have not should be easily understood, since they follow the same general rules.

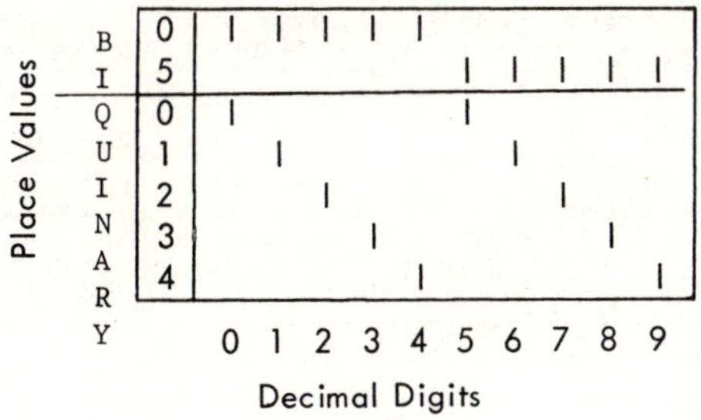


Figure 3-22. *The biquinary code.*

### Problems

1. What is a bit?
2. How is a bit represented in paper tape? magnetic tape? punched cards? computer core?
3. What advantages has an eight-bit code over a five-bit code?
4. What is parity? How is it used?
5. What are some of the weaknesses of the baudot code?
6. In what ways do the hollerith and Remington Rand card codes differ?
7. Define "packed decimal." What would be the advantage(s) of using packed decimal in a data communications system?
8. What is BCD?
9. Convert $8274_{10}$ to its BCD equivalent.
10. What qualities should a good communications code possess?
11. What is "code transparency?" How is it used?

## TABLE 3-3 COMMUNICATION CODES

| Col. No. | 1 | 2 | 3 | 4 | 5 |
|---|---|---|---|---|---|
| Code | Baudot (Telegraphic) | CCITT#2 (International) | ASCII (With even parity) | EBCDIC (Extended BCD Interchange Code) | Field Data (6-bit) |
| No. Bits | 5 | 5 | 7 | 8 | 6 |
| Parity | None | None | Even | None | None |
| Media | Tape Channels | Tape Channels | Tape Channels | Tape Channels | Tape Channels |
| A | 1-2 | 1-2 | 1-7 | 1-7-8 | 2-3 |
| B | 1-4-5 | 1-4-5 | 2-7 | 2-7-8 | 1-2-3 |
| C | 2-3-4 | 2-3-4 | 1-2-7-8 | 1-2-7-8 | 4 |
| D | 1-4 | 1-4 | 3-7 | 3-7-8 | 1-4 |
| E | 1 | 1 | 1-3-7-8 | 1-3-7-8 | 2-4 |
| F | 1-3-4 | 1-3-4 | 2-3-7-8 | 2-3-7-8 | 1-2-4 |
| G | 2-4-5 | 2-4-5 | 1-2-3-7 | 1-2-3-7-8 | 3-4 |
| H | 3-5 | 3-5 | 4-7 | 4-7-8 | 1-3-4 |
| I | 2-3 | 2-3 | 1-4-7-8 | 1-4-7-8 | 2-3-4 |
| J | 1-2-4 | 1-2-4 | 2-4-7-8 | 1-5-7-8 | 1-2-3-4 |
| K | 1-2-3-4 | 1-2-3-4 | 1-2-4-7 | 2-5-7-8 | 5 |
| L | 2-5 | 2-5 | 3-4-7-8 | 1-2-5-7-8 | 1-5 |
| M | 3-4-5 | 3-4-5 | 1-3-4-7 | 3-5-7-8 | 2-5 |
| N | 3-4 | 3-4 | 2-3-4-7 | 1-3-5-7-8 | 1-2-5 |
| O | 4-5 | 4-5 | 1-2-3-4-7-8 | 2-3-5-7-8 | 3-5 |
| P | 2-3-5 | 2-3-5 | 5-7 | 1-2-3-5-7-8 | 1-3-5 |
| Q | 1-2-3-5 | 1-2-3-5 | 1-5-7-8 | 4-5-7-8 | 2-3-5 |
| R | 2-4 | 2-4 | 2-5-7-8 | 1-4-5-7-8 | 1-2-3-5 |
| S | 1-3 | 1-3 | 1-2-5-7 | 2-6-7-8 | 4-5 |
| T | 5 | 5 | 3-5-7-8 | 1-2-6-7-8 | 1-4-5 |
| U | 1-2-3 | 1-2-3 | 1-3-5-7 | 3-6-7-8 | 2-4-5 |
| V | 2-3-4-5 | 2-3-4-5 | 2-3-5-7 | 1-3-6-7-8 | 1-2-4-5 |
| W | 1-2-5 | 1-2-5 | 1-2-3-5-7-8 | 2-3-6-7-8 | 3-4-5 |
| X | 1-3-4-5 | 1-3-4-5 | 4-5-7-8 | 1-2-3-6-7-8 | 1-3-4-5 |
| Y | 1-3-5 | 1-3-5 | 1-4-5-7 | 4-6-7-8 | 2-3-4-5 |
| Z | 1-5 | 1-5 | 2-4-5-7 | 1-4-6-7-8 | 1-2-3-4-5 |
| 0 | *P | *P | 5-6 | 5-6-7-8 | 5-6 |
| 1 | *Q | *Q | 1-5-6-8 | 1-5-6-7-8 | 1-5-6 |
| 2 | *W | *W | 2-5-6-8 | 2-5-6-7-8 | 2-5-6 |
| 3 | *E | *E | 1-2-5-6 | 1-2-5-6-7-8 | 1-2-5-6 |
| 4 | *R | *R | 3-5-6-8 | 3-5-6-7-8 | 3-5-6 |
| 5 | *T | *T | 1-3-5-6 | 1-3-5-6-7-8 | 1-3-5-6 |

| | | | | | |
|---|---|---|---|---|---|
| 6 | *Y | *Y | 2-3-5-6 | 2-3-5-6-7-8 | 2-3-5-6 |
| 7 | *U | *U | 1-2-3-5-6-8 | 1-2-3-5-6-7-8 | 1-2-3-5-6 |
| 8 | *I | *I | 4-5-6-8 | 4-5-6-7-8 | 4-5-6 |
| 9 | *O | *O | 1-4-5-6 | 1-4-5-6-7-8 | 1-4-5-6 |
| U CASE | 1-2-4-5 | 1-2-4-5 | | 2-3-5-6 | 1 |
| L CASE | 1-2-3-4-5 | 1-2-3-4-5 | | 2-3 | 2 |
| SPACE | 3 | 3 | 6-8 | 7 | 1-3 |
| CAR RET | 4 | 4 | 1-3-4-8 | | 3 |
| L FEED | 2 | 2 | 2-4 | 1-3-6 | 1-2 |
| BELL | *S | *J | 1-2-3-8 | | |
| # | *H | | 1-2-6-8 | 1-2-4-5-6-7 | |
| $ | *D | | 3-6 | 1-2-4-5-7 | 1-2-3-6 |
| % | | | 1-3-6-8 | 3-4-6-7 | |
| & | *G | | 2-3-6-8 | 5-7 | |
| ' | *J | | 1-2-3-6 | 1-3-4-5-6-7 | |
| ( | *K | *L | 4-6 | 1-3-4-7 | 1-4-6 |
| ) | *L | *L | 1-4-6-8 | 1-3-4-5-7 | 6 |
| * | | | 2-4-6-8 | 3-4-5-7 | 4-6 |
| + | | *Z | 1-2-4-6 | 2-3-4-7 | 2-6 |
| , | *N | *N | 3-4-6-8 | 1-2-4-6-7 | 2-3-4-6 |
| - | *A | *A | 1-3-4-6 | 6-7 | 1-6 |
| . | *M | *M | 2-3-4-6 | 1-2-4-7 | 1-3-4-5-6 |
| / | *X | *X | 1-2-3-4-6-8 | 1-6-7 | 3-4-5-6 |
| : | *C | | 2-4-5-8 | 2-4-5-6-7 | 1-2-4-6 |
| ; | *V | | 1-2-4-5 | 2-3-4-5-7 | 1-2-4-5-6 |
| < | | | 3-4-5-8 | 5-7-8 | 1-2-6 |
| = | | *V | 1-3-4-5 | 2-3-4-5-6-7 | 3-6 |
| > | | | 2-3-4-5 | 7-8 | 1-3-6 |
| ? | *B | *B | 1-2-3-4-5-8 | 2-4-7 | 3-4-6 |
| @ | | | 7-8 | 3-4-5-6-7 | |
| ⌑ | | | | | |
| ! | *F | | 1-6 | 2-4-5-7 | 1-3-4-6 |
| " | *Z | | 2-6 | 1-4-5-6-7 | 2-4-6 |
| ¢ | | | | 1-2-3-4-5-7 | |
| [ | | | 1-2-4-5-7-8 | | |
| \ | | | 3-4-5-7 | | |
| ] | | | 1-3-4-5-7-8 | | |
| ← | | | 1-2-3-4-5-7 | 3-4-7 | |
| ↑ | | | 2-3-4-5-7-8 | | |
| TAB | | | | 1-3 | |
| PCH ON | | | | 3-5-6 | |

*Upper case

**TABLE 3-3. COMMUNICATION CODES (cont.)**

| Col. No. | 1 | 2 | 3 | 4 | 5 |
|---|---|---|---|---|---|
| Code | Baudot (Telegraphic) | CCITT#2 (International) | ASCII (With even parity) | EBCDIC (Extended BCD Interchange Code) | Field Data (6-bit) |
| No. Bits | 5 | 5 | 7 | 8 | 6 |
| Parity | None | None | Even | None | None |
| Media | Tape Channels | Tape Channels | Tape Channels | Tape Channels | Tape Channels |
| PCH OFF | | | | 3 | |
| DELETE | | | | 1-2-3 | |
| NULL | | | | | |
| SOM | | | | | |
| EOA | | | | | |
| EOM | | | | | |
| EOT | | | | 1-2-3-5-6 | |
| WRU | | *D | | | |
| RU | | | | | |
| ALT MODE | | | | | |
| STOP | | | | 1-3-5-6 | 1-2-3-4-6 |
| SPECIAL | | | | | 2-3-4-5-6 |
| IDLE | | | | 1-2-3-5 | 1-2-3-4-5-6 |

| Col. No. | 7 | 8 | 9 | 10 | 11 |
|---|---|---|---|---|---|
| Code | Mach - 10 (Chan 5 is for odd parity) | Flexowriter Model SFD | IBM Perforated Tape and Transmission Code | IBM Card Code (Hollerith) | UNIVAC Card Code (90 - Col.) |
| No. Bits | 6 | 6 | 7 | | |
| Parity | 5th level odd | 5th level odd | 5th level odd | | |
| Media | Tape Channels | Tape Channels | Tape Channels | Card Row | Card Row |
| A | 1-6-7 | 1-6-7 | 1-6-7 | 12-1 | 1-5-9 |
| B | 2-6-7 | 2-6-7 | 2-6-7 | 12-2 | 1-5 |
| C | 1-2-5-6-7 | 1-2-5-6-7 | 1-2-5-6-7 | 12-3 | 0-7 |
| D | 3-6-7 | 3-6-7 | 3-6-7 | 12-4 | 0-3-5 |
| E | 1-3-5-6-7 | 1-3-5-6-7 | 1-3-5-6-7 | 12-5 | 0-3 |
| F | 2-3-5-6-7 | 2-3-5-6-7 | 2-3-5-6-7 | 12-6 | 1-7-9 |
| G | 1-2-3-6-7 | 1-2-3-6-7 | 1-2-3-6-7 | 12-7 | 5-7 |
| H | 5-6-7 | 4-6-7 | 4-6-7 | 12-8 | 3-7 |
| I | 1-4-5-6-7 | 1-4-5-6-7 | 1-4-5-6-7 | 12-9 | 3-5 |

| | | | | | |
|---|---|---|---|---|---|
| J | 1-5-7 | 1-5-7 | 1-5-7 | 11-1 | 1-3-5 |
| K | 2-5-7 | 2-5-7 | 2-5-7 | 11-2 | 3-5-9 |
| L | 1-2-7 | 1-2-7 | 1-2-7 | 11-3 | 0-9 |
| M | 3-5-7 | 3-5-7 | 3-5-7 | 11-4 | 0-5 |
| N | 1-3-7 | 1-3-7 | 1-3-7 | 11-5 | 0-5-9 |
| O | 2-3-7 | 2-3-7 | 2-3-7 | 11-6 | 1-3 |
| P | 1-2-3-5-7 | 1-2-3-5-7 | 1-2-3-5-7 | 11-7 | 1-3-7 |
| Q | 4-5-7 | 4-5-7 | 4-5-7 | 11-8 | 3-5-7 |
| R | 1-4-7 | 1-4-7 | 1-4-7 | 11-9 | 1-7 |
| S | 2-5-6 | 2-5-6 | 2-5-6 | 0-2 | 1-5-7 |
| T | 1-2-6 | 1-2-6 | 1-2-6 | 0-3 | 3-7-9 |
| U | 3-5-6 | 3-5-6 | 3-5-6 | 0-4 | 0-5-7 |
| V | 1-3-6 | 1-3-6 | 1-3-6 | 0-5 | 0-3-9 |
| W | 2-3-6 | 2-3-6 | 2-3-6 | 0-6 | 0-3-7 |
| X | 1-2-3-5-6 | 1-2-3-5-6 | 1-2-3-5-6 | 0-7 | 0-7-9 |
| Y | 4-5-6 | 4-5-6 | 4-5-6 | 0-8 | 1-3-9 |
| Z | 1-4-6 | 1-4-6 | 1-4-6 | 0-9 | 5-7-9 |
| 0 | 6 | 6 | 2-4-5 | 0 | 0 |
| 1 | 1 | 1 | 1 | 1 | 1 |
| 2 | 2 | 2 | 2 | 2 | 1-9 |
| 3 | 1-2-5 | 1-2-5 | 1-2-5 | 3 | 3 |
| 4 | 3 | 3 | 3 | 4 | 3-9 |
| 5 | 1-3-5 | 1-3-5 | 1-3-5 | 5 | 5 |
| 6 | 2-3-5 | 2-3-5 | 2-3-5 | 6 | 5-9 |
| 7 | 1-2-3 | 1-2-3 | 1-2-3 | 7 | 7 |
| 8 | 4 | 4 | 4 | 8 | 7-9 |
| 9 | 1-4-5 | 1-4-5 | 1-4-5 | 9 | 9 |
| U CASE | 3-4-5-6-7 | 3-4-5-6-7 | 2-3-4 | 9-6 | |
| L CASE | 2-4-5-6-7 | 2-4-5-6-7 | 2-3-4-6-7 | 12-9-6 | |
| SPACE | 5 | 5 | 5 | Blank | Blank |
| CAR RET | 8 | 8 | 1-3-4-5-7 | 11-9-5 (cr-lf) | |
| L FEED | | | 1-3-4-5-6 | 0-9-5 | |
| BELL | | | | | |
| # | 1-2-3-4-5 | *3 | 1-2-4 | 3-8 | 0-1-5-7 |
| $ | 1-2-4-5-7 | *4 | 1-2-4-5-7 | 11-3-8 | 0-1-3-5-9 |
| % | *V | *5 | *5 | *5 | 0-1-5 |
| & | 5-6-7 | *7 | 5-6-7 | 12 | 0-1-3-5-7 |
| ' | *X | *6 | *6 | *6 | 0-1-5-7-9 |
| ( | *D | *9 | *9 | *9 | 0-1-9 |
| ) | *M | *0 (Zero) | *0 (Zero) | *0 (Zero) | 0-13-5 |
| * | 1-3-4-5-6 | *8 | | *8 | 0-1 |

*Upper case

## TABLE 3-3. COMMUNICATION CODES (cont.)

| Col. No. | 7 | 8 | 9 | 10 | 11 |
|---|---|---|---|---|---|
| Code | Mach - 10 (Chan 5 is for odd parity) | Flexowriter Model SFD | IBM Perforated Tape and Transmission Code | IBM Card Code (Hollerith) | UNIVAC Card Code (90 - Col.) |
| No. Bits | 6 | 6 | 7 | | |
| Parity | 5th level odd | 5th level odd | 5th level odd | | |
| Media | Tape Channels | Tape Channels | Tape Channels | Card Row | Card Row |
| + | | | *& | *& | 1-5-7-9 |
| , | 1-2-4-5-6 | 1-2-4-5-6 | 1-2-4-5-6 | 0-3-8 | 0-3-5-9 |
| - | 7 | 7 | 7 | 11 | 0-3-5-7 |
| . | 1-2-4-6-7 | 1-2-4-6-7 | 1-2-4-6-7 | 12-3-8 | 1-3-5-9 |
| / | 1-5-6 | 1-5-6 | 1-5-6 | 0-1 | 3-5-7-9 |
| : | *4 | *; | *4 | *4 | 1-3-7-9 |
| ; | *P | 5-6-7 | *3 | *3 | 1-3-5-7-9 |
| < | *G | | | | 0-1-5-9 |
| = | *U | | *1 | *1 | 0-1-3-5-7-9 |
| > | *7 | | | | 0-3-5-7-9 |
| ? | | */ | */ | */ | 0-1-3 |
| @ | *5 | *2 | 6 | 4-8 | 0-1-3-7 |
| ¤ | *E | | *2 | *2 | 0-1-3-9 |
| ! | | | *$ | | 0-3-7-9 |
| " | *W | *- | *7 | | |
| ¢ | * * | | *@ | *@ | |
| [ | | | | | 0-5-7-9 |
| \ | | | | | 0-1-3-7-9 |
| ] | | | | | 1-3-5-7 |
| ← | | | | | |
| ↑ | | | | | |
| TAB | 2-3-4-5-6 | 2-3-4-5-6 | 1-3-4-6-7 | 12-5-9 | |
| PCH ON | 3-4-7 | 3-4-7 | 3-4-5 | 4-9 | |
| PCH OFF | 1-2-3-4-6 | 1-2-3-4-6 | 3-4-5-6-7 | 12-4-9 | |
| DELETE | | | 1-2-3-4-5-6-7 | 12-7-9 | |
| NULL | | | | | |
| SOM | | | | | |
| EOA | | | | | |
| EOM | | | | | |
| EOT | | | 1-2-3-4-5 | 7-9 | |
| WRU | | | | | |
| RU | | | | | |

| | | | | | |
|---|---|---|---|---|---|
| ALT MODE | | | | | |
| STOP | 1-2-4 | 1-2-4 | | | |
| SPECIAL | | | | | |
| IDLE | | | 1-2-3-4-7 | 11-7-9 | |
| BYPASS | | | 3-4-6 | 0-4-9 | |
| RESTORE | | | 3-4-7 | 11-4-9 | |
| ± | | | 1-2-4 | *# | |
| ‡ | | | 2-4-6 | 0-2-8 | |
| ǂ | | | | *‡ | |
| γ | | | | *–0 (Minus 0) | |
| √ | | | | *+0 | |
| RDR STOP | | | 1-3-4 | 5-9 | |
| EOB | | | 2-3-4-5-6 | 0-6-9 | |
| BACKSP | | | 2-3-4-5-7 | 11-6-9 | |
| +0 | | | 2-4-5-6-7 | 12-0 | |
| –0 | | | 2-4-7 | 11-0 | |
| PREFIX | | | 1-2-3-4-6 | | |
| CANCEL | | | 5-7 | | |

| Col. No. | 12 | 13 | 14 | 15 | 16 |
|---|---|---|---|---|---|
| Code | SEVEN - BIT Alphameric (Mod. BCD) | SIX - BIT Numeric | XS - 3 (Excess 3) | BI-QUINARY | 2 - OUT - OF - 5 |
| No. Bits | | 6 | 6 | 7 | 5 |
| Parity | | None | None | Self-cont. even | Self-cont. even |
| Media | C BA 8421 | C F 8421 | B A 8421 | 05 01234 | 0 1 2 3 6 |
| A | 1 11 0001 | | 0 1 0100 | | |
| B | 1 11 0010 | | 0 1 0101 | | |
| C | 0 11 0011 | | 0 1 0110 | | |
| D | 1 11 0100 | | 0 1 0111 | | |
| E | 0 11 0101 | | 0 1 1000 | | |
| F | 0 11 0110 | | 0 1 1001 | | |
| G | 1 11 0111 | | 0 1 1010 | | |
| H | 1 11 1000 | | 0 1 1011 | | |
| I | 0 11 1001 | | 0 1 1100 | | |
| J | 0 10 0001 | | 1 0 0100 | | |
| K | 0 10 0010 | | 1 0 0101 | | |
| L | 0 10 0011 | | 1 0 0110 | | |
| M | 0 10 0100 | | 1 0 0111 | | |

*Upper case

## TABLE 3-3. COMMUNICATION CODES (cont.)

| Col. No. | 12 | 13 | 14 | 15 | 16 |
|---|---|---|---|---|---|
| Code | SEVEN - BIT Alphameric (Mod. BCD) | SIX - BIT Numeric | XS - 3 (Excess 3) | BI-QUINARY | 2 - OUT - OF - 5 |
| No. Bits | | 6 | 6 | 7 | 5 |
| Parity | | None | None | Self-cont. even | Self-cont. even |
| Media | C BA 8421 | C F 8421 | B A 8421 | 05 01234 | 0 1 2 3 6 |
| N | 1 10 0101 | | 1 0 1000 | | |
| O | 1 10 0110 | | 1 0 1001 | | |
| P | 0 10 0111 | | 1 0 1010 | | |
| Q | 0 10 1000 | | 1 0 1011 | | |
| R | 1 10 1001 | | 1 0 1100 | | |
| S | 0 01 0010 | | 1 1 0101 | | |
| T | 1 01 0011 | | 1 1 0110 | | |
| U | 0 01 0100 | | 1 1 0111 | | |
| V | 1 01 0101 | | 1 1 1000 | | |
| W | 1 01 0110 | | 1 1 1001 | | |
| X | 0 01 0111 | | 1 1 1010 | | |
| Y | 0 01 1000 | | 1 1 1011 | | |
| Z | 1 01 1001 | | 1 1 1100 | | |
| 0 | 0 00 1010 | 1 0 0000 | 0 0 0011 | 10 10000 | 0 1 1 0 0 |
| 1 | 1 00 0001 | 0 0 0001 | 0 0 0100 | 10 01000 | 1 1 0 0 0 |
| 2 | 1 00 0010 | 0 0 0010 | 0 0 0101 | 10 00100 | 1 0 1 0 0 |
| 3 | 0 00 0011 | 1 0 0011 | 0 0 0110 | 10 00010 | 1 0 0 1 0 |
| 4 | 1 00 0100 | 0 0 0100 | 0 0 0111 | 10 00001 | 0 1 0 1 0 |
| 5 | 0 00 0101 | 1 0 0101 | 0 0 1000 | 01 10000 | 0 0 1 1 0 |
| 6 | 0 00 0110 | 1 0 0110 | 0 0 1001 | 01 01000 | 1 0 0 0 1 |
| 7 | 1 00 0111 | 0 0 0111 | 0 0 1010 | 01 00100 | 0 1 0 0 1 |
| 8 | 1 00 1000 | 0 0 1000 | 0 0 1011 | 01 00010 | 0 0 1 0 1 |
| 9 | 0 00 1001 | 1 0 1001 | 0 0 1100 | 01 00001 | 0 0 0 1 1 |
| U CASE | | | | | |
| L CASE | | | | | |
| SPACE | 1 01 0000 | | 0 0 0000 | | |
| CAR RET | | | | | |
| L FEED | | | | | |
| BELL | | | | | |
| # | 1 00 1011 | | 0 1 1101 | | |
| $ | 0 10 1011 | | 1 0 0010 | | |
| % | 1 01 1100 | | 1 1 0001 | | |
| & | 0 11 0000 | | 0 1 0000 | | |
| | | | 1 0 0000 | | |

| | | | | | |
|---|---|---|---|---|---|
| ( | | | 1 0 1101 | | |
| ) | | | 1 1 1111 | | |
| * | 1 10 1100 | | 1 0 0001 | | |
| + | | | 1 1 0011 | | |
| , | 0 01 1011 | | 1 1 0010 | | |
| - | 1 10 0000 | | 0 0 0010 | | |
| . | 1 11 1011 | | 0 1 0010 | | |
| / | 0 01 0001 | | 1 1 0100 | | |
| : | | | 0 1 0001 | | |
| ; | | | 0 0 1110 | | |
| < | | | 0 1 1110 | | |
| = | | | 0 1 1111 | | |
| > | | | 1 1 1110 | | |
| ? | | | 0 1 0011 | | |
| @ | 0 00 1100 | | 1 0 1110 | | |
| ⌑ | 0 11 1100 | | 1 1 1101 | | |
| ! | | | 1 0 0011 | | |
| " | | | | | |
| ¢ | | | | | |
| [ | | | | | |
| \ | | | | | |
| ] | | | | | |
| ← | | | | | |
| ↑ | | | | | |
| TAB | | | | | |
| PCH ON | | | | | |
| PCH OFF | | | | | |
| DELETE | | | | | |
| NULL | | | | | |
| SOM | | | | | |
| EOA | | | | | |
| EOM | | | | | |
| EOT | | | | | |
| WRU | | | | | |
| RU | | | | | |
| ALT MODE | | | | | |
| STOP | | | | | |
| SPECIAL | | | | | |
| IDLE | | | | | |
| GROUP MK. | 0 11 1111 | | | | |
| TAPE MK. | 0 00 1111 | | | | |
| DRUM MK. | 0 00 0000 | | | | |
| +0 | 0 11 1010 | | | | |
| −0 | 1 10 1010 | | | | |

*Upper case

*CHAPTER 4*

# Characteristics of Transmission Media

## INTRODUCTION

It is axiomatic that the most important part of a communications system is the channel. Its bandwidth limits the volume of information that may be transmitted within certain time limits. Any imperfections, such as interference and distortions, that may be present also affect the maximum possible volume of information and its accuracy; in addition, they contribute to the complexity of the terminal devices. A channel's cost is a major factor in the design and use of a communications system.

The balance of this discussion will cover the characteristics (bandwidth, interference, distractions, distortion, and net loss) of communications channels in general, and their effects on information passed through these channels. Because most telephone circuits were originally designed for voice transmission, much of the terminology and technology is a holdover from voice transmission.

## BANDWIDTH

### Frequency Defined

Electrical currents can be classified in two general categories — direct or alternating. Direct current (d c) travels in only one direction in a circuit, while alternating current (a c) travels first in one direction (+), then in the other (−).

Figure 4-1 shows graphically an alternating current starting at a zero value, going through its positive phase and returning to zero, then going through its negative phase and again returning to zero. This is one cycle of alternating current. Either half of the cycle is called a *baud*.

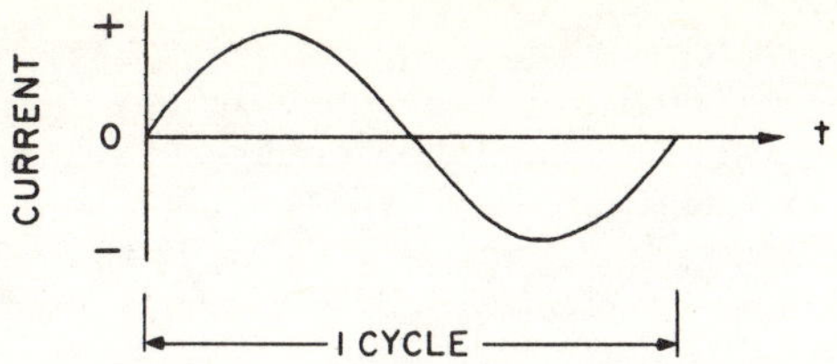


Figure 4-1. *One cycle of alternating current.*

The number of cycles completed per second is the current's *frequency*, and is expressed in hertz.[1]

## The Frequency Spectrum

Frequencies vary over an extremely wide range, beginning at zero and increasing progressively through acoustics, radio, infrared (heat), light, ultraviolet, X-rays, gamma rays, and cosmic rays. The acoustical range is from about 20 Hz to about 20,000 Hz, and varies considerably from person to person. The radio range extends from about 14 kHz to over 10 million kHz.

Figure 4-2 indicates the disparity between the frequencies that can be detected by the human ear and those that can be transmitted over a telephone channel. The human voice (100 to 1100 Hz), however, falls mostly within the limits imposed by the telephone circuit.

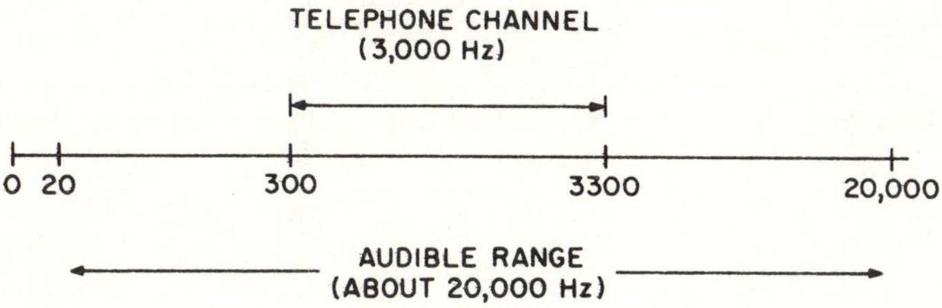


Figure 4-2. *Frequency disparity between human ear and telephone channel.*

## Frequency of a Data Signal

A data communication signal often consists of a *range* of frequencies. The frequency of a signal at any particular moment depends on the composition of the code being transmitted. For illustration, assume a character with the binary representation of 11110000 is transmitted. If binary 1's are given a positive voltage and 0's a negative voltage, only one cycle

[1]The term hertz, abbreviated Hz, has taken the place of cycles per second to promote international understanding.

would have been transmitted during the time required to transmit the character: the line voltage would have gone from zero to a positive voltage (for the duration of the 1 bits) and then would have swung down through zero to a negative voltage while the 0's were transmitted.

On the other hand, if a character with the binary equivalent of 10101010 were transmitted, four cycles of current would have occurred during the same character-time. In fact, transmission of the second character would have resulted in the highest possible frequency for that particular signal, since it had caused the greatest number of transitions from one signal state (positive) to the other (negative). Thus the number of bits a transmission channel can carry per unit of time is directly related to the upper limit of its usable frequency range.

## Bandwidth and Passband

Bandwidth is a measurement of the width of a range of frequencies. The telephone channel described in the above graph has a bandwidth of 3000 Hz (3 kHz).

A passband, on the other hand, is a slot at a certain place in the frequency spectrum which allows a particular range of frequencies (bandwidth) to pass. The passband of the telephone channel was defined by its limits of 300 and 3300 Hz. Notice that passband defines a *particular slot* in the frequency spectrum, while bandwidth defines a *range* of frequencies.

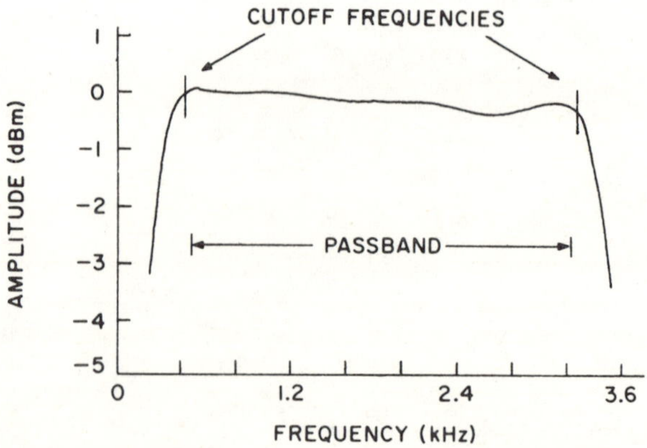


Figure 4-3. *Passband created by using filters.*

A major difference between the grades of available channels (teletypewriter, voice, television, etc.) is their bandwidth. Bandwidth has a great deal to do with the quality of a received signal as compared to the signal originally transmitted. Intelligibility of the human voice, for example, requires a bandwidth of about 400 Hz. Articulation — the pitch and tonal qualities of a voice — requires about 1200 Hz of bandwidth, three times that of intelligibility.

### Cutoff Frequencies

Since communications media often have many simultaneous conversations (or other information) imposed upon them, it is necessary to restrict each conversation to its own path. The electrical filters that are used for this purpose create a passband which allows frequencies within a certain range to pass through the circuit, but will block all frequencies outside this range. The points at the upper and lower edges of the passband are called cutoff frequencies. (See Fig. 4-3.)

## DISTORTION

### Introduction

If it were possible to transmit a signal over a channel that had no imperfections, the signal would arrive at its destination exactly as it had been sent. Perfect channels, however, exist only in theory, thus signals become distorted during transmission.

It has been pointed out earlier that noise is an unpredictable phenomenon that is best described statistically. Distortion, however, has a fixed affect on a signal, and is a function of each individual channel. There are three types of distortion that a channel may impart to a signal: delay distortion, attenuation distortion, and jitter.

### Delay (Phase) Distortion

To have some degree of understanding about the effects a channel imparts on signals it carries, it is helpful to examine the characteristics of a perfect signal, and compare these with the distorted signal which appears at the receiving end of a channel.

Signals can be constructed in many ways. In the following discussion, signals composed of alternating currents of various frequencies will be discussed. First, a simple single-frequency wave is shown, then, a complex two-frequency wave, and finally, the effects of channel distortion will be examined.

### Simple and Complex Waves

A simple (single-frequency) wave is shown in Fig. 4-4. The simple wave (signal) shown here can be represented by a rotating vector. Vectors of this type are assumed to rotate counterclockwise, with the reference point, zero

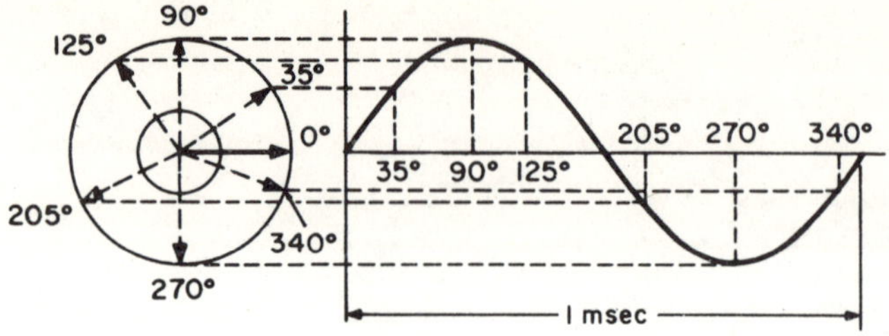


Figure 4-4. *Representation of a simple wave.*

degrees, at the right. One complete rotation of the vector represents one complete cycle of the signal. Since the waveform shows that one cycle of the signal is completed in 1 millisecond, the signal frequency is 1000 Hz. Therefore, the vector will rotate 1000 times per second.

Complex waves result from combining two or more simple waves. The instantaneous value of a complex wave is the sum of the instantaneous values of the simple waves it comprises. Figure 4-5 represents a complex wave (A + B) and its two components (A and B) as the signal was transmitted and as it would be received if sent through a perfect channel. Affects of channel imperfections on signals are discussed in following paragraphs.

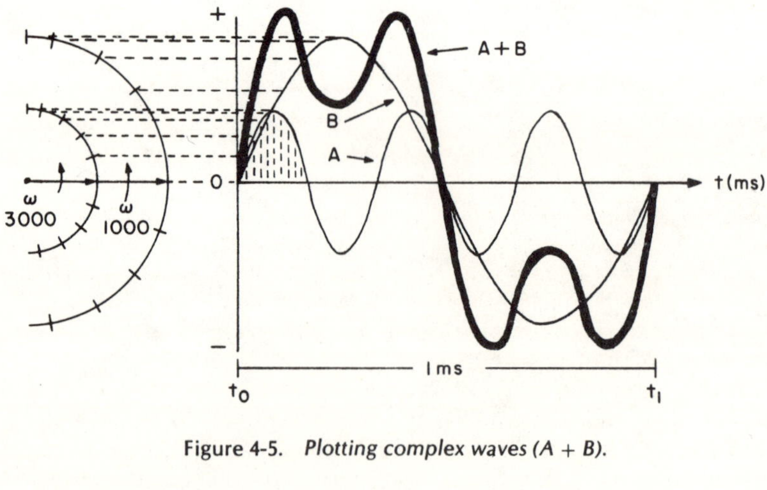


Figure 4-5. *Plotting complex waves (A + B).*

## Propagation Time

A signal's propagation time is a function of both the nature of the channel through which it is traveling and the frequency of the signal itself. A channel with no external influences and no resistance would move a signal at a maximum of 186,000 miles per second (the speed of light). A microwave carrier might move the signal at 100,000 miles per second (0.01 ms per mile), and a pair of wires in a cable would pass the signal at about 14,000 miles per second (0.0714 ms per mile). An additional 1.2 ms is used whenever the signal passes through terminals which convert from carrier to wire (or vice versa), filters, and other equipment.

If the channel is distortionless, all frequencies will pass through it at the same speed. Under these circumstances, the frequency and the phase of any given signal will have a constant (linear) relationship with respect to time (see Fig. 4-6). The interval of time between the instant that a signal is transmitted and the instant it is received is called phase delay, absolute delay, or propagation time. All signals have some phase delay, but the mere fact that a signal arrives later than it was sent poses few problems in data transmission.

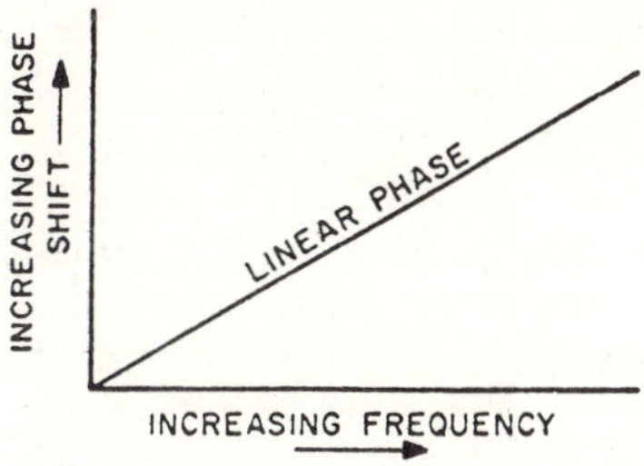


Figure 4-6. *Phase delay through a distortionless channel.*

## Envelope Delay (Phase Delay Distortion)

A cause of difficulty in data transmission is that the shift of phase with respect to frequency is not usually linear in most transmission media, but has a curve similar to that of Fig. 4-7. Under these circumstances, some frequencies of a complex signal will be delayed more than others during transmission, resulting in a distortion of the original signal. This phenomenon is called envelope delay, phase delay distortion, or merely phase distortion.

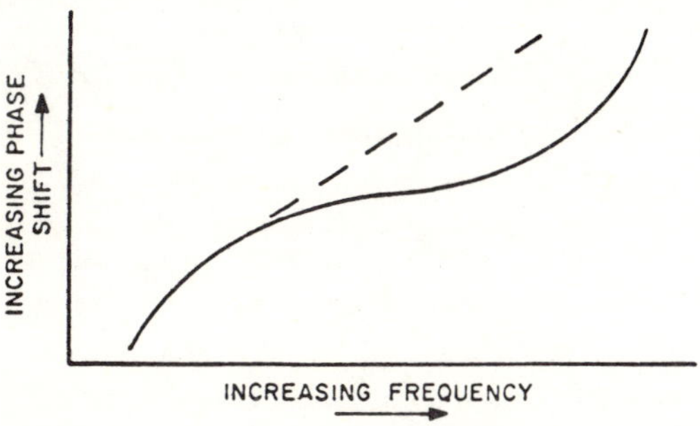


Figure 4-7. *Phase delay of normal voice channel.*

Mathematically, envelope delay is defined as the first derivative of phase delay. This means that the shape of the envelope delay curve reflects the degree of *change* of the slope of the phase vs frequency curve. A linear phase shift vs frequency (dotted line of Fig. 4-7) results in no envelope delay (dotted line of Fig. 4-8), while a nonlinear phase shift vs frequency results in a distortion of the transmitted signal.

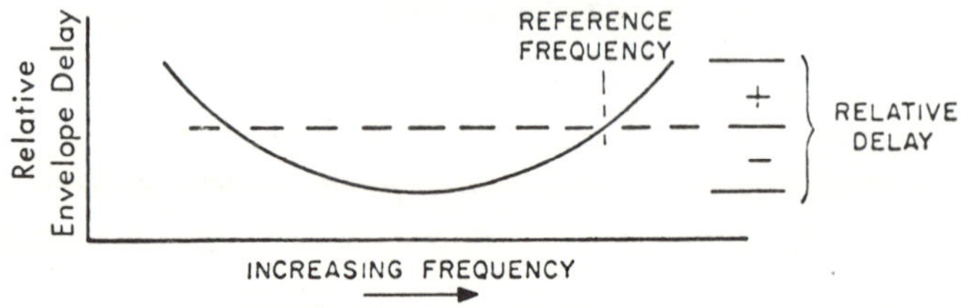


Figure 4-8. *Relative envelope delay curve.*

If we were to transmit a binary signal at 1000 bauds, each signal element (baud) would have a duration of 1 ms. A relative envelope delay of only 1 ms between the mark and space frequencies, then, would cause the two frequencies to be superimposed at the receiver, obliterating the signal.

To understand the effects of envelope delay, compare the following three diagrams. The first, Fig. 4-9, is the same as Fig. 4-4, and shows the perfect complex waveform as it was transmitted at t = 0. The 1000Hz and 3000Hz components are in phase at t = 0.

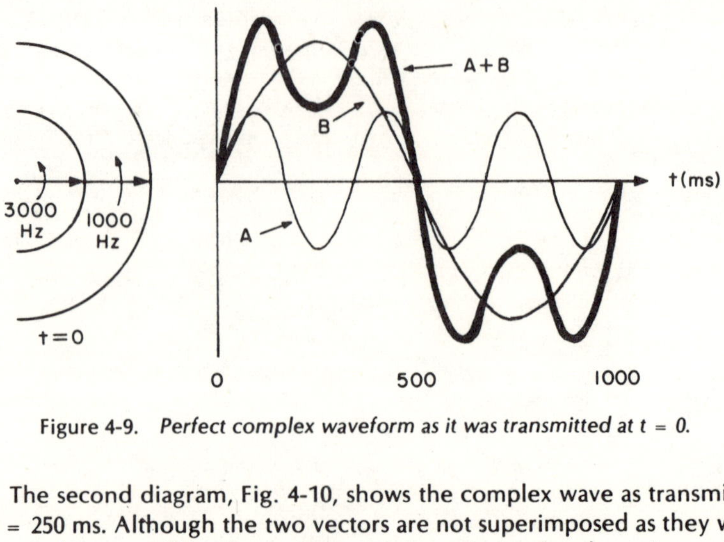


Figure 4-9. *Perfect complex waveform as it was transmitted at t = 0.*

The second diagram, Fig. 4-10, shows the complex wave as transmitted at t = 250 ms. Although the two vectors are not superimposed as they were at t = 0, they are still in phase, since one vector is rotating three times as fast

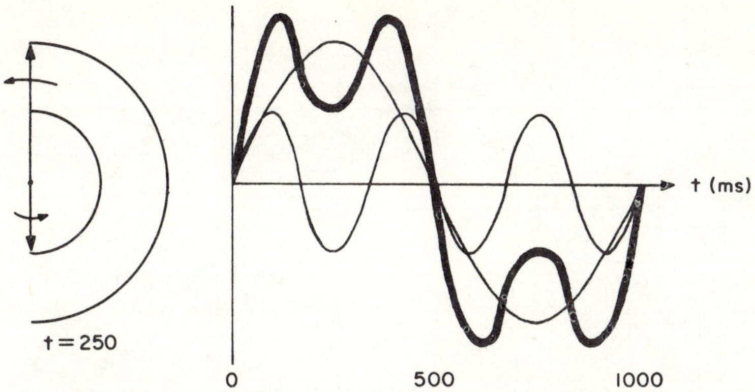


Figure 4-10. *Complex waveform as it was transmitted at t = 250.*

as the other. As long as they maintain this 3-to-1 relationship, they will be in phase with each other.

The third diagram, Fig. 4-11, shows what happens to a complex waveform when its components fall out of phase. The transmitted signal was the same as above, but the channel has delayed the low-frequency component, thus changing the resulting waveform.

### Delay Equalizers

Although envelope delay will always be present in communications circuits, it is important to flatten this delay across the frequency bandwidth of the channel to minimize delay distortion. This is done by adding "delay

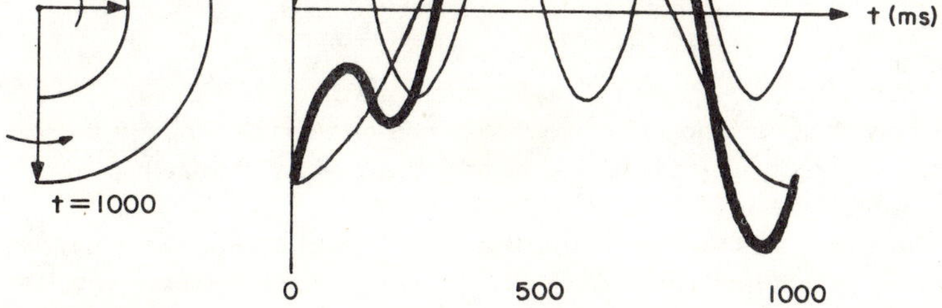


Figure 4-11. *Complex waveform with its components out of phase.*

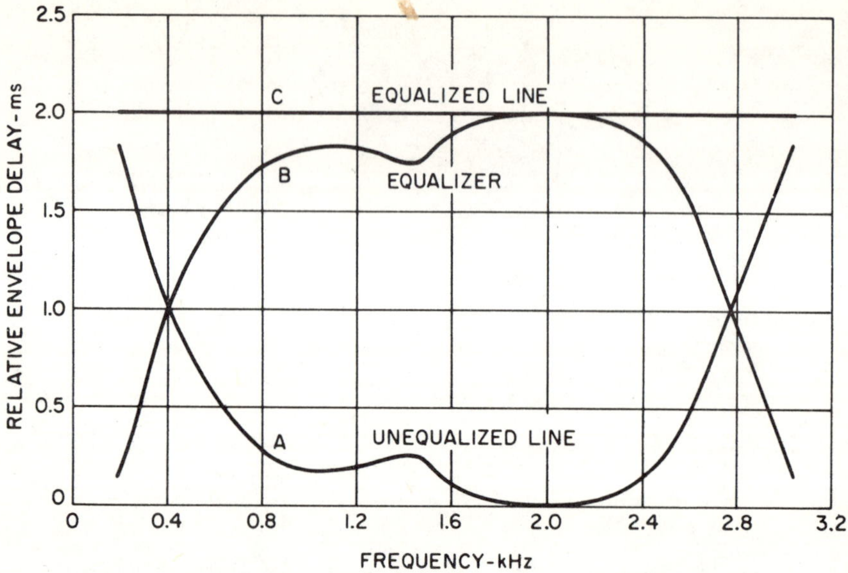


Figure 4-12. *Flattening envelope delay by using delay equalizers. (Courtesy of Rixon Electronics)*

equalizer" networks to the channel circuits (see Fig. 4-12). The equalizers introduce a delay which is inverse to that of the channel alone, and the cumulative result is a relatively flat delay across the channel's bandwidth.

## DISTRACTIONS

### Distractions Defined

The term "distractions" is usually used to define extraneous audible signals which appear on the line during a telephone conversation. Distractions consist of crosstalk, echo, or both.

### Crosstalk

Crosstalk is the "slopping over" of the contents of one telephone circuit to another circuit. This causes voices and music, which normally should not be present, to be audible in the second circuit.

The cause of crosstalk is the induction of currents by one circuit to physically adjacent circuits. Many circuits, composed of a pair of wires, pass through cables which may contain over two thousand such pairs, but which are not more than 2½ in. in diameter! This proximity often provides enough inductive force to cause crosstalk. Factors which tend to increase the effects of crosstalk include the frequency and the strength of the source currents and the distance that the circuits parallel each other.

**Echo**

Echo is the return of your own voice during a telephone conversation. It usually happens only on long-distance calls and is caused by an electrical imbalance of the circuit.

If A is talking, and any of his speech energy is looped back toward him by an unbalanced electrical network at B, then A will hear his own voice as an echo. (see Fig. 4-13) A contributing factor is that most present circuits carry voice signals at less than 20,000 miles per second. If the distance and the speed of the circuit are such that any echo returns in more than 45 ms, then it will be noticed.

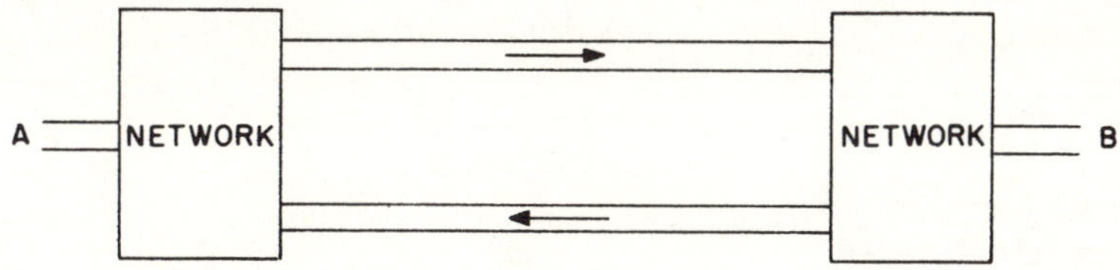


Figure 4-13. *Simplified telephone circuit.*

**Echo Suppressors**

Echo suppressors were developed to eliminate this distraction, and they are in wide use today. As A begins to talk, his voice energy activates a relay which short-circuits the return path, thus blocking any echo. When A stops talking, the echo suppressor is quickly deactivated (10 ms) and B can begin talking (see Fig. 4-14). This operation is the cause of the phenomenon you may have observed during a long-distance phone call, where all line noise

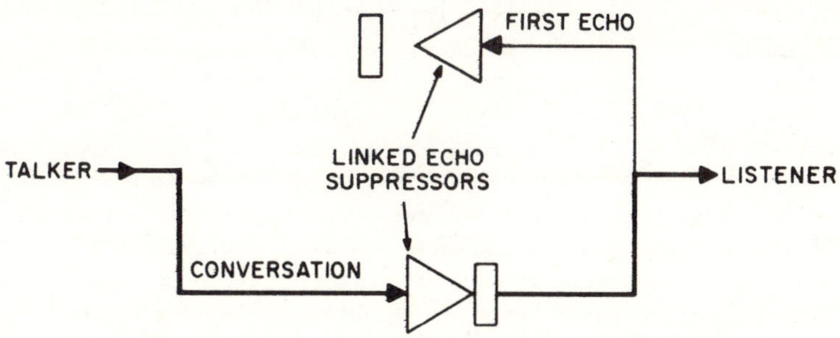


Figure 4-14. *Principle of echo suppression. (Courtesy of IBM)*

stopped when you started talking, and you seemed to be talking to yourself. You may also have noticed that the distant party had tried to start talking before you were through, but you couldn't hear him until you stopped talking and the line seemed to "open up" again.

### Disablers

Echo suppressors work well for what they are intended to do, but they cause serious problems for data transmission. First, any business machine or data set using voice circuits equipped with echo suppressors must allow up to a second for turn-around time — the time it takes to reverse transmission directions. Second, and more important in many cases, is that if an echo suppressor is activated by a transmission from machine A, any attempt by machine B to send back an interrupt signal to A will be blocked. Disablers deactivate echo suppressors by applying a 2025 Hz tone to the line for about 300 ms (± 50 ms) when no other signals are being transmitted. Any interval of 100 ms or more will reactivate the echo suppressors.

## INTERFERENCE

### Interference Defined

While distractions are annoying to people during telephone conversations, they may or may not have an adverse affect on data transmission. Interference, on the other hand, often is a source of errors during data transmissions and can arise from induction, noise, or multifrequency tones.

### Induction

Both magnetic and electrical induction can cause interference in parallel circuits. This is particularly true as the volume of data being transmitted increases, since data is applied more continuously to circuits than is voice. What difference does this make? Quite a bit, since the amount of induction depends largely on the cumulative strength of any signals present.

Suppose, for example, that a cable carried a voice conversation on each of its 100 pairs of wires and that each voice was well within required strength limitations — caused no induction, in other words. There are gaps in voice conversation, so that at any instant there may have been no more than 75 voices, and a tolerable amount of noise, actually present within the 100-pair cable. The 75 voices produced a limited and harmless inductive force.

Now let's substitute data transmissions for each of the 100 voice conversations. The gaps between words have been eliminated, so that now we have continuous energy eminating from the equivalent of nearly 100 circuits, rather than the equivalent of the 75 or so we had before. If there are electrical or magnetic imbalances within this group of circuits or if one or more circuits contain signals or noise of excessive power, the entire group of circuits can be affected to the extent that they are unable to carry error-free data.

## NOISE

Channel noise consists of random electrical impulses. These unwanted signals are introduced by a variety of sources, and are generally classified as either impulse noise or white noise. Noise is considered as interference when it reaches the proportion that it causes errors in transmission.

### Impulse Noise

Impulse noise is usually caused by the operation of machinery and switches, and by electrical storms. It is characterized by its short but intense duration and its confinement to a limited portion of the frequency spectrum. Within the audio range it can be heard as sharp clicks or bursts of static. (See Fig. 4-15.)

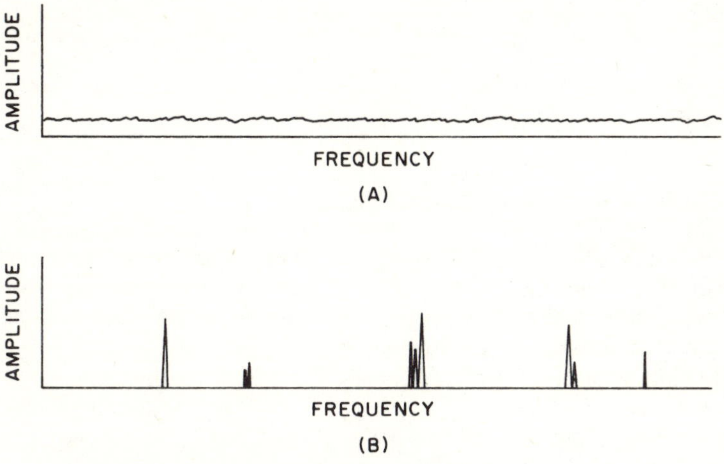


Figure 4-15. *Amplitude and frequency distribution of (A) white noise, and (B) impulse noise.*

### White Noise (Gaussian Noise)

White noise, on the other hand, has its energy spread out over a broad range of the frequency spectrum and is heard as the familiar background hiss on the radio or telephone. Its causes include powerline induction, cross-modulation from adjacent circuits, and a conglomeration of other random signals. One explanation for using "white" to describe this type of noise is that it causes the snowlike phenomenon seen on TV when the signal is weak.

## Noise as an Error Source

Noise becomes bothersome when it exceeds a magnitude of about half that of a positive code element. This is because samples are taken of a signal, and if noise exceeds the decision level, the noise is interpreted as a positive signal (see Fig. 4-16).

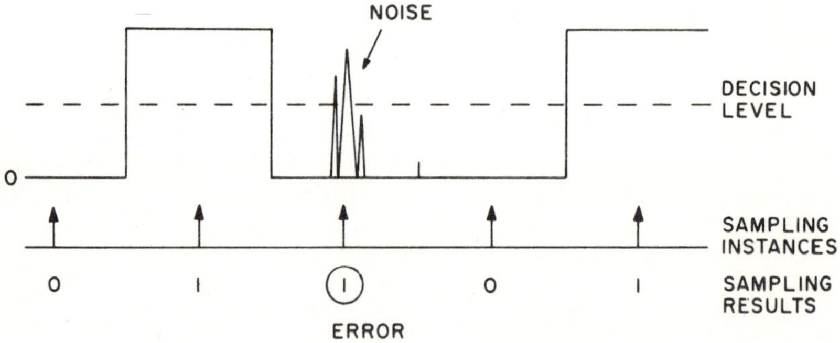


Figure 4-16. *Effect of noise on a binary signal.*

## Effect of Noise on Channel Capacity (Shannon)

Since the unwanted signals that are noise have many of the same characteristics as an information-carrying signal, we must find some way of creating a clear distinction between the two. Fortunately, the power level (intensity) of noise is quite low on most circuits. If the power of the information signal is considerably above that of the noise, the receiving equipment can more easily distinguish between them. As the signal and the noise approach the same power level, while channel bandwidth remains constant, the signal has to exist for longer periods of time in each of its discrete conditions or states to enable the receiving equipment to distinguish it from the random states of the noise.

C. E. Shannon did some pioneering work in this area in 1949 and developed a theory stating that the theoretical maximum bit rate, C, through a channel of bandwidth BW and signal-to-random-noise-power ratio of S/N (where S = signal power and N = noise power) is given by this formula:

$$C = BW \log_2 (1 + S/N)$$

The S/N power ratio indicates the relative strength of the signal to that of the channel noise. It is expressed either in ratio form ($10^3$: 1) or decibels (dB). For example, an S/N power ratio of $10^3$: 1 also could be expressed as 30 dB; a ratio of $10^2$ : 1 would equate to 20 dB, and so on.

If we had a perfect channel with a 3000-Hz bandwidth and an S/N power ratio of $10^3$: 1, we could use the above formula and calculate the maximum bit rate of the channel:

$$\begin{aligned} C &= BW \log_2(1 + S/N) \\ &= 3000 \log_2(1 + 10^3) \\ &= 3000 \log_2(1001) \\ &= 3000 \times 10 \text{ (approx.)} \\ &= 30{,}000 \text{ bits per second (approx.)} \end{aligned}$$

Note that the coding and modulation methods are not described; they may be nearly impossible to achieve, and certainly would not be economical.

### Noise Penalty for Multilevel Code Elements

In the presence of noise, a binary signal is more easily and accurately detected than one using several bits per code element. As the bit content (number of levels) of a code element is increased, a corresponding increase in the S/N power ratio must be made to maintain equal detection results relative to a binary signal. The above formula can be modified to give the required minimum S/N power ratio from a known bit rate and bandwidth.

$$S/N = 2^{C/BW}$$

Applying this formula to binary and multilevel signals will show the extent of the noise penalty required to permit the transmission of various multilevel signals.

The S/N power ratio for a binary signal must first be found to serve as a reference. Assuming a perfect 3000-Hz channel, Nyquist's rate of 6000 bps may be used, with the result that a minimum S/N power ratio of 3:1 is required:

$$\begin{aligned} S/N &= 2^{C/BW} - 1 \\ S/N &= 2^{6000/3000} - 1 = 2^2 - 1 = 3 \end{aligned}$$

The decibel equivalent of a 3:1 S/N power ratio is:

$$\begin{aligned} dB &= 10 \log S/N \\ &= 10 \log 3 = 10\,(.48) = 4.8 \end{aligned}$$

In contrast to the binary system above, a ternary (three-level) system would require a higher S/N power ratio. The maximum bit rate of a ternary system through the ideal 3000-Hz channel is:

$$\begin{aligned} bps &= 2\,BW (\log_2 3) \\ &= 6000\,(1.58) = 9500 \end{aligned}$$

and the required S/N power ratio is:

$$
\begin{aligned}
S/N &= 2^{C/BW} - 1 \\
&= 2^{9500/3000} - 1 \\
&\cong 2^3 - 1 = 7 \text{ (approx.)}
\end{aligned}
$$

The decibel equivalent of a S/N power ratio of 7 is:

$$dB = 10 \log 7 = 8.5$$

The noise penalty of a ternary system relative to a binary system (in an ideal channel) is thus 8.5 − 4.8 = 3.7 dB. A quaternary system requires a minimum difference of 11.7 dB between the signal and noise power levels, and thus has a noise penalty of 11.7 − 4.8 = 6.9 dB above binary. (These are minimum requirements of an otherwise perfect channel and are shown here to indicate the extent of the noise penalty that is required to increase signal speed through a given channel.)

In addition to the limit that bandwidth and channel noise (reduced signal-to-noise-power ratio) impose on the bit-carrying capacity of a given channel, other channel imperfections, and limitations of present equipment, impose a practical minimum S/N power ratio in the range of $10^2$: 1 (20 dB) or more.

### Multifrequency Tones

A relative newcomer to the classification of interference is a group of multifrequency tones. These tones, sometimes heard during telephone conversations, sound somewhat like musical horns on an automobile. The source of the tones varies from Touch-Tone dialers on telephones and teletypewriters to data-set signals, and their presence on a circuit is becoming more prevalent. Since tones are the language of data equipment, the presence of unwanted tones caused by induction is sometimes misinterpreted as valid data by various business machines or data sets which converse in these tones.

## ATTENUATION

### Definition

In communications, attenuation refers to the loss of power that a signal suffers as it travels from the transmitter to a receiver. In other words, it is the power that is absorbed by the transmission medium.

### Measurement

One of the most practical ways to measure attenuation is by applying logarithms to the ratio of the input (transmitted) power to the output (received) power. The common logarithm of a number is merely how many times 10 must be multiplied by itself to give that number. The logarithm of 100, for example, is 2, since 10 must be multiplied by itself twice: $10 \times 10 = 100$. The logarithm of one million is 6: $10 \times 10 \times 10 \times 10 \times 10 \times 10 = 1{,}000{,}000$. In the shorthand of mathematics, this would be written: $\log 1{,}000{,}000 = 6$.

Since $\log 10 = 1$ and $\log 100 = 2$, it follows that log 47 will be a value somewhere between 1 and 2. (One would probably guess that it might be about 1.5, which is close, because it's 1.672.)

## Decibels

The units which result from taking the logarithm of the *ratio* of input power to output power are called bels; one-tenth of a bel is a decibel. If an amplifier produced a power output, $P_O$, of 100 watts from an input, $P_I$, of only 1 watt, then its gain would be the results of dividing $P_I$ into $P_O$, or a ratio of

$$P_O / P_I = 100/1 \text{ or } 100$$

This amplification can also be expressed by the logarithm of the ratio: $\log 100 = 2$.

The result above would be called a gain of 2 bels, or 20 decibels (20 dB for short). Thus, a decibel equals 10 times the logarithm of the result of dividing the power output by the power input:

$$dB = 10 \log P_O / P_I$$

One further point: A decibel, alone, does not have an inherent value; it merely indicates the relationship between two degrees of power. Thus, a 10 dB loss gives a good idea of the attenuation of a signal over a circuit, but it gives no indication of the original strength of the signal.

Assume a signal being transmitted with 1.2 mW (milliwatts) of energy, and a channel that absorbs .6 mW. What is the dB loss of the signal as it passes through the channel?

Answer:

$$\begin{aligned} dB &= 10 \log P_O / P_I \\ &= 10 \log 0.6/1.2 \\ &= 10 \log 0.5 \\ &= 10\,(-0.3) = -3 \\ &= 3 \text{ dB loss} \end{aligned}$$

## Attenuation Distortion

High frequencies lose strength more rapidly than low frequencies during transmission through a medium, thus a received signal can be distorted by an unequal attenuation, or loss, of its component frequencies. In the example in Fig. 4-17, the high frequency has experienced more loss of strength through attenuation than the low-frequency portion of the signal.

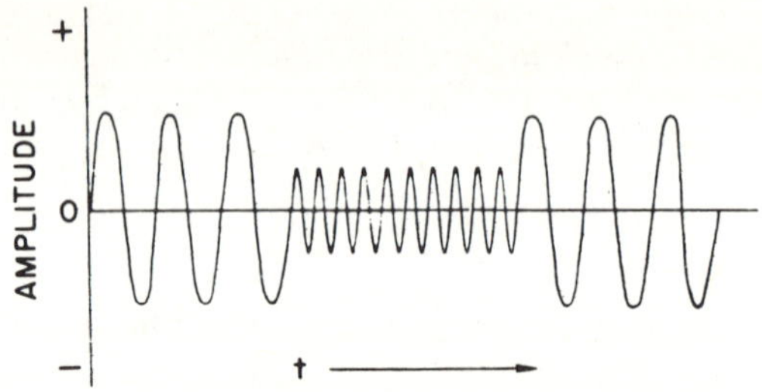


Figure 4-17. *Effect of attenuation distortion.*

To overcome this attenuation distortion, attenuation equalizers are used. These are electrical networks which have frequency losses complimentary to those of the line. Thus, when added to the line circuit, they give a net result of equalizing the loss of all frequency components of a signal. (See Fig. 4-18.)

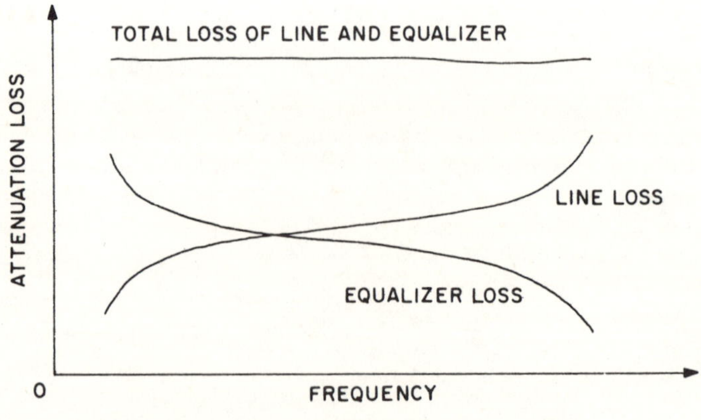


Figure 4-18. *Attenuation distortion correction.*

## NET LOSS

### Definition

Net loss is merely the difference in power of a signal between its source and its destination, and is measured in dB.

### dBm

Because the decibel, as a measure of the ratio of two powers, has no intrinsic value, it often would be helpful to have a known value to use as a reference to express the power level of a signal. The Bell System developed such a standard: 1 mW of power at 1000 Hz into a 600-ohm resistance impedance. This is called "zero dBm." If a signal has more power than this standard, it has plus dBm and if it has less than this standard power, it has minus dBm.

### dBrn

A 1000-Hz tone with 1 mW of power would cause interference in a voice or a data circuit. To reduce this noise (interference) to a negligible level would require lowering it to about –90 dBm. Rather than refer to noise levels in negative terms (such as –40 dBm) the –90 dBm level was taken as a standard, and measurements made in reference to this standard are in terms of dBrn (dB above reference noise).

It should be evident by now that some sort of reference should always be given with an expression using dB. "A 10-dB gain" has meaning, whereas "a 40-dB signal" lacks definition.

## JITTER

### Definition

Since two frequencies are used in most cases to transmit a binary signal, the circuit characteristics will affect each in a different way. The combined effects of both attenuation and delay distortion results in peak distortion, or jitter. Jitter, then, is the variation in time of the received sequence of tone transitions as compared to the time sequence in which they were transmitted.

### Discussion

If mark and space frequencies are both from a portion of the passband that has relatively flat attenuation and delay characteristics, jitter will be at a minimum. If the frequencies were affected similarly to those in Fig. 4-11, however, jitter would probably be quite noticeable, or even troublesome.

To realize how jitter can become so much of a problem, consider that the function of the receiving terminal is to recreate squarewave pulses which duplicate those that were transmitted by the originating business machine. To do this, it must watch for zero crossings (transitions) in the incoming signal. As the level of the incoming signal crosses zero as it goes from a positive to a negative value, the output of the receiving terminal

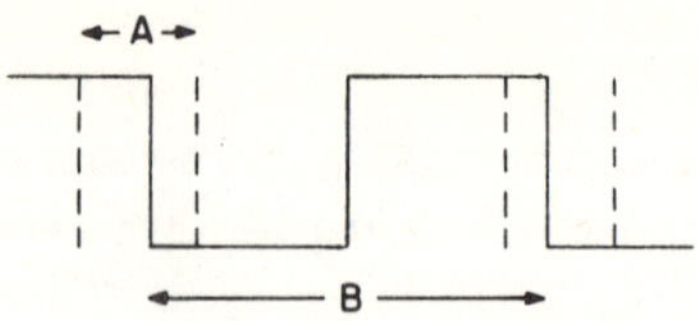


Figure 4-19. *Calculation of percent jitter. (Bell System Monograph 3580)*

changes instantaneously to a negative d-c value. As long as the received signal stays on the negative side of zero, the output will be a steady negative d-c value. When the incoming signal again crosses zero in its transition to a positive state, the terminal will convert to a steady positive output.

If the transmitted signal becomes distorted during its course through the channel, the timing of its zero crossings at the receiving end will not correspond to those when it was transmitted. As the severity of the channel distortion changes, the zero crossings of the output of the receiving terminal will also fluctuate, causing jitter in the resultant squarewave signal. If the receiving terminal is deriving its timing from the received signal, or if the sampling is out of phase with the transmitter, errors in the interpretation of the signal by the receiving terminal are likely.

Jitter is measured by dividing A, the variation in received time of the transitions between frequencies, as compared to the transmitted time sequence, by B, the time required to transmit two bits (see Fig. 4-19). Under these terms the maximum possible distortion is 50 percent.

$$\text{Percent jitter} = \frac{\text{Max. var. in transition time}}{\text{Time of two bits}} \times 100$$

$$= \frac{A}{2\,B} \times 100$$

**Problems**

1. What are the characteristics of a channel?
2. Define frequency.
3. What is the usual frequency range (upper and lower frequencies) of the human voice? the human ear? a telephone voice channel?
4. Compare the terms "bandwidth" and "passband."
5. Name and define two distractions.
6. Name two kinds of noise, and give the causes of each.
7. What effect does noise have on channel capacity?
8. Name two kinds of distortion and describe their effect on a signal.
9. What is the function of a delay equalizer?
10. What is jitter? Why is it undesirable?

*CHAPTER 5*

# *Efficiency and Error Control*

## INTRODUCTION

People working in communications are always faced with the challenge of pumping the highest possible volume of accurate data through the most economical systems available. The closer we come to this goal, the higher the resulting efficiency and the lower the costs.

In general, the efficiency of a communications system is affected by: (1) Characteristics of the circuit (bandwidth, distortion, noise, attenuation, etc.); (2) echo suppressor turn-around time (on the switched network); (3) coding schemes (baudot, hollerith, etc.); (4) speed of the transmitting and receiving equipment ("effective" speeds); (5) error control schemes.

The following discussion will cover the last two of these points; the first three were covered in previous chapters.

Efficiency is determined by comparing the number of information bits that are moved through a system in a given time interval to the maximum number it would be possible to move. The discussion of efficiency will be introduced by an explanation of redundancy and its role in data transmission.

## REDUNDANCY

### Definition

Redundant information is superfluous, or unnecessary, to the meaning of the original data. Examples of redundancy are parity bits, hash totals, and check digits, since they do not amplify the meaning of the original data. The use of parallel channels, each carrying the same information, and the retransmission of messages are also examples of redundancy.

Redundancy is added to original data to help reduce the chances of undetected errors occurring during the transmission or processing of data. If enough redundancy is added to original data, it is possible to devise codes which have the capability of not only detecting errors, but correcting them as well.

If any form of error control is to be used, it is necessary to add redundancy to the coded characters of the message. All five bits of the baudot code are required to describe any one particular character, so baudot does not contain any redundancy or any means of error control. To clarify this concept, consider the reasoning that a teletypewriter might use if it could think during the receipt of a coded character. Table 5-1 assumes that the machine is in upper case:

**TABLE 5-1.**

| Received Bit | Possible Characters | | | | | | | | | | | | | | | |
|---|---|---|---|---|---|---|---|---|---|---|---|---|---|---|---|---|
| 1 = Mark | – | ? | $ | 3 | ! | ' | ( | 1 | BELL | 7 | 2 | / | 6 | " | LTRS | FIGS |
| 2 = Space | | ? | $ | 3 | ! | | | | BELL | | | / | 6 | " | | |
| 3 = Mark | | | | | ! | | | | BELL | | | / | 6 | | | |
| 4 = Space | | | | | | | | | BELL | | | | 6 | | | |
| 5 = Space | | | | | | | | | BELL | | | | | | | |

Each time another bit arrived, the number of remaining characters was halved. When the final bit arrived, the machine caused the bell to ring, but there was no way to know with certainty that the transmitted character was the bell character. If the baudot code used a sixth bit for parity, there would have been a somewhat greater degree of assurance that a "bell" character had actually been sent. But then a degree of redundancy to the original five-bit code would have been introduced.

## Calculation

Redundancy is calculated with this formula:

$$\text{Redundancy} = 1 - \frac{\text{Information Bits}}{\text{Total Bits}}$$

A seven-bit code using an eighth bit for parity would have a redundancy of:

$$R = 1 - (\text{Bits}_I/\text{Bits}_T)$$
$$= 1 - 7/8 = 1 - 0.875 = 0.125 \text{ or } 12.5\%$$

# EFFICIENCY

## Coding Schemes

Partially to overcome the lack of error control in codes such as baudot and hollerith, codes with built-in aids toward error control were devised.

These codes require that a proportionately larger number of bits per character be transmitted, and thus make less efficient use of available circuit time.

The efficiency of any coding system is measured by dividing the information bits by the total bits:

$$\text{Efficiency} = \frac{\text{Information Bits}}{\text{Total Bits}} = \frac{\text{Bits}_I}{\text{Bits}_T}$$

Using this formula, we find that the baudot code is extremely efficient:

$$\begin{aligned} E &= \text{Bits}_I/\text{Bits}_T \\ &= 5/5 = 1 \text{ or } 100\% \end{aligned}$$

But a seven-bit code using an eighth bit for parity has a lower efficiency:

$$E = \text{Bits}_I/\text{Bits}_T = 7/8 = 87.5\%$$

Since only four bits are required to send numerical information ($2^4 = 16 > 10$) the above eight-bit code would be very inefficient if a system were limited to sending only numbers:

$$E = \text{Bits}_I/\text{Bits}_T = 4/8 = 50\%$$

As an aside, efficiency can also be derived from a known redundancy.

$$\text{Efficiency} = 1 - \text{Redundancy}$$

If the redundancy of 12.5% from the previous example is used, then:

$$\begin{aligned} E &= 1 - R \\ &= 1 - 0.125 = 0.875 \text{ or } 87.5\% \end{aligned}$$

**Transmission Systems**

In addition to the loss of communication efficiency inherent in coding schemes, there is an efficiency loss due to the methods used to transmit the data. Teletypewriters, for example, add 2.42 pulses to the five pulses of every baudot character, giving a maximum system efficiency of 67.4 percent:

$$\begin{aligned} E &= \frac{\text{Pulses}_I}{\text{Pulses}_T} \\ &= 5/7.42 = 67.4\% \end{aligned}$$

Thus, during transmission of baudot code over a teletypewriter system, better than 30 per cent of the available time is taken up by the start and stop pulses alone.

### Combined Coding and Transmission

A high price, therefore, is paid to send information safely through a communications system. First, possible information bits are often used to provide for a parity coding scheme, then another percent of the possible maximum is given up to allow synchronization of the sending and receiving machines.

If a system is rated at 1200 bps, how many true information bits can it be expected to carry? Assume an ASCII code and a paper tape reader which uses one start and one stop pulse for each character.

ASCII is a seven-bit code which adds an eighth bit for even parity. Its efficiency:

$$E = \text{Bits}_I/\text{Bits}_T$$
$$= 7/8 = 87.5\%$$

The efficiency of the transmission device:

$$E = \text{Pulses}_I/\text{Pulses}_T$$
$$= 8/10 = 80\%$$

The efficiency of the total system ($E_s$) is found by multiplying the coding efficiency ($E_c$) by the transmission efficiency ($E_t$):

$$E_s = (E_c)(E_t)$$
$$= (87.5)(80) = 70\%$$

So the maximum information transfer of the system rated at 1200 bps is really (1200) (0.70) = 840 bps.

The remaining pulses are not information bits, but redundancy and control pulses required for error control and operation of the equipment. Properly, this should have been described as an 840 bps system at 1200 *bauds.* The 840 bps was the system's *effective* speed. The following paragraphs give more detail on effective speeds.

## EFFECTIVE SPEEDS

### Operator and Machine Capabilities

Many machines and systems are described in terms of the maximum words per minute (wpm), characters per second (cps), or bits per second (bps) that they will handle. It is foolish, however, to design an operation based on these maximum speeds. An automatic (tape transmitting) teletypewriter, for example, which is rated at 60 wpm actually delivers not more than 51 wpm under normal circumstances. This lower effective speed is due to the use of the functions, such as the upper and lower case shifts, the line feeds, and the carriage returns, that are an inherent part of most messages. If the same machine is operated manually, its effective speed should be assumed to be no more than 40 wpm. Table 5-2 indicates the rated and effective speeds for several teletypewriter systems (half duplex).

TABLE 5-2. RATED VS EFFECTIVE SPEEDS

| Rated Speed (WPM) | Type of Transmission | Maximum Effective Speed (WPM) |
|---|---|---|
| All | Manual | 40 |
| 60 | Automatic | 51 |
| 75 | " | 64 |
| 100 | " | 85 |

## Rated vs Effective Speeds

As shown above, the effective speed of a system is seldom the speed at which it is rated. It is important, then, to calculate the effective speed of any system contemplated for use.

Assume, for example, that a machine salesman claims that his paper-tape transmitter is rated at a speed of 1050 bps. What is its effective speed?

The code used is the five-bit baudot, so at 1050 bps a speed of 210 cps could be assumed. However, the punch used in the receiver operates at 105 characters per second. As a result the effective speed is only 105 cps, instead of 210. The efficiency?

$$5 \text{ bits/char} \times 105 \text{ char/sec} = 525 \text{ bps}$$

Then:

$$E = \text{Bits}_I/\text{Bits}_T = 525/1050 = 50\%$$

The other 525 bits comprised block characters, synchronization characters, and bits used to pad the five-bit characters (the machine actually transmits an eight-bit character plus control characters).

If the same machine were to transmit an eight-channel code, as it is capable of doing at the same speed of 105 cps, its effective speed would remain at the 105 cps speed of the punch. Its efficiency, however, would increase:

$$8 \text{ bits/char} \times 105 \text{ char/sec} = 840 \text{ bps}$$

Then:

$$E = \text{Bits}_I/\text{Bits}_T = 840/1050 = 80\%$$

### Effective Speeds and Block Length

The overall efficiency of a communication system can be affected significantly by its message block length. A block that is too short produces an adverse ratio of transmission time to turn-around time. Exceedingly long blocks increase the probability that an error will be included and require excessive retransmission time. It is important, therefore, to determine the optimum block length, keeping in mind the consideration of other system factors, such as available buffer sizes, coding schemes, error control methods, and the like.

## INCREASING CHANNEL CAPACITY

### Bauds and Bandwidth (Nyquist)

In 1928, H. Nyquist showed that one cycle of a signal can contain a maximum of two bauds (See Fig. 2-7). Thus, the maximum theoretical signalling speed (in bauds) of a circuit is twice its bandwidth:

$$\text{Bauds} = 2\,\text{BW}$$

If, for example, a channel had a bandwidth of 3000 Hz, its theoretical maximum signaling speed would be 6000 bauds. This is called signaling at the "Nyquist rate" and could be achieved only over a perfect channel, free from noise and distortion.

Charges for transmission channels are a large portion of the cost of a data communications system, so it is understandable that effort should be made toward making the most efficient use of available facilities.

Since the number of bauds that can be sent over any particular channel is limited by the channel's characteristics, each baud should be made to carry as much information as possible. In straight binary code, each baud carries no more than one bit. In ternary, quaternary, or even more sophisticated coding schemes, however, the bit content of each baud can be increased, thus increasing the bit rate without increasing the signaling speed. One way to do this is to use multilevel code elements.

### Code Elements

A code element is the elemental unit from which a code is constructed. Marks (1's) and spaces (0's) are code elements of the baudot telegraphic code, for example. A baud may contain only one code element, but the code element may contain more than one bit if a sophisticated coding scheme is used.

A binary system would assign each code element one of two possible values, so its information content would be one bit. A ternary code element can have any of three possible values, a quaternary can have four, and so forth. A general way to state this is that any N-ary code element can have any one of N possible values. Assuming that both the value of the code element and its sequential position are completely random, the amount of information in a message can be measured by multiplying the number of code elements, m, times the logarithm of N:

$$I = m \log_2 N$$

If a message consisted of five quaternary code elements, for example, the information transferred would be:

$$I = m \log_2 N$$
$$= 5 \log_2 4 = 5 \times 2 = 10 \text{ bits}$$

**Dibits**

One of the simplest methods of increasing the rate of bits through a channel is to combine pairs of adjacent bits into dibits (pronounced "die bits"), and let each baud carry one dibit. For example, the bit sequence shown in Fig. 5-1 is divided into pairs, or dibits.

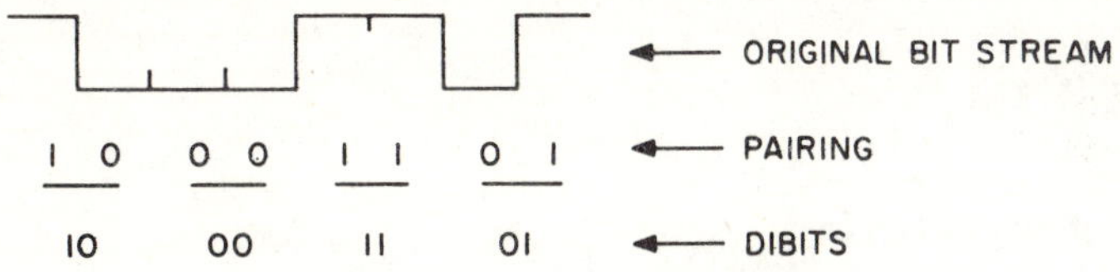


Figure 5-1. *Pairing of bits into dibits.*

These dibits are the only four combinations that bits can assume if taken two at a time. The objective is to send each dibit as one multilevel signal element, or baud. If converted, a different voltage is assigned to each of the four combinations; then each dibit is converted to its respective voltage before sending it out on the channel, and bits will be transmitted at twice the channel's normal capacity.

If the four possible dibit combinations are assigned these voltages:

| Dibit | Voltage |
|---|---|
| 11 | 3 |
| 10 | 2 |
| 01 | 1 |
| 00 | 0 |

then the same sequence of bits given above can be converted to dibits, which, in turn, are presented to the channel as variable voltages (See Fig. 5-2).

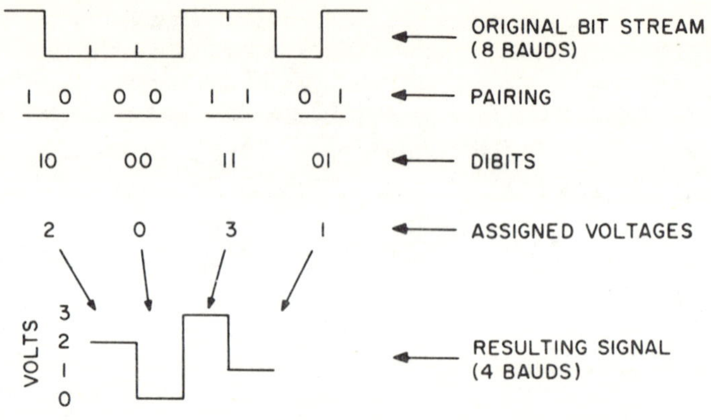


Figure 5-2. *Dibits presented to channel as variable voltages.*

Sent at the same baud rate as before, these eight bits required only half the previous amount of time. Instead of voltages, four *tones* of different frequencies could have been used. In fact, tones are used extensively in data transmission. Being analog signals, tones have the advantages of being able to pass through the dialed telephone network and of being less susceptible to attenuation (fading) and noise. The time gained by using dibit code elements isn't free, however, because these are no longer the simple and easily detected bit. Terminal equipment has had to be designed that can discriminate between four levels of voltages or tones and also perform translation between these signals and the bits they carry.

Another method of preparing dibits for transmission, used by the Bell System's 201A and 201B data sets, is to convert each dibit to a shift in phase of the transmitted signal. Since there are four distinct dibits, each can be related to one of four phases of an analog signal. The 00 dibit, for example, might be represented by a phase value of 90°, and the 01, 10, and 11 dibits by phase values of 180°, 270°, and 0° respectively. This form of modulation is discussed in more depth in the next chapter.

## ERROR CONTROL

Any degree of effort could be made to assure that messages arrive at their destinations unchanged. If a telegram or TWX message were sent, for example, only the minimum of precautionary measures would have been taken, since there would be no assurance that errors had not occurred during transmission. If sufficient requirements existed for more favorable transmission reliability, a private line could be leased from a common carrier, or the speed of transmission could be reduced; the leased line could be conditioned to have fewer sources of error-causing disturbances, and lower speeds also may reduce errors by allowing the detection equipment more time to sample the state of the incoming signal. Complete assurance that the message would be error-free when received would still be lacking, however.

A step which would give some degree of the assurance wanted would be to add error detection to the system. If most of the errors were detected and tagged, their location would be known, but this still would not indicate, at the receiving end, what the message *should* have been. And error detection systems often fail to tag all errors. To be sure, within an extremely small margin of error, that the received message was absolutely correct, some method of error detection and *correction* (EDC) should be employed.

Several methods of EDC, some more reliable than others, are described below. Each has its own philosophy of reducing to a minimum the effects of error-causing disturbances.

## ERROR PREVENTION

### The Operator

Supposedly, it should be unnecessary to mention operator errors. Even though the machine operator is a major source of errors in data communications, the field of human engineering will not be discussed in this book.

### The Machine

Many machines have been designed to prevent many of the errors that operators often make. This does not necessarily mean that the machines will prevent a "6" key from being depressed when a "9" key would have been correct. It does mean that machines such as Addo-X, Add-punch, C-Dek, and others have self-contained programs that confine the operator to a set sequence of operations, thus assuring to some degree that format and message sequence will be correct.

Some machines put an initial and final sequence on the machine's tape output to permit its direct placement on transmission systems. Others add parity bits or characters to the data originally entered, thus lending the machine output to use in EDC systems.

### The Facilities

As mentioned above, ordinary telegraph and telephone lines can be used for data transmission with good results, provided that reasonable precautions are taken. "Good results" should be recognized as a relative term; a few errors on an administrative teletypewriter network may have no adverse effects, whereas the same number of errors may be unacceptable for computer data processing.

No two communications links will have identical characteristics, but all have conditions that cause errors in data transmission. Even the experts disagree, however, on the pattern in which these conditions are found. In 1959, Alexander, Gryb, and Nast of the Bell Telephone System, for example, made an exhaustive study of the capacity of the telephone exchange network to carry data communications. Among other things, they concluded that 70 percent of all errors were isolated one-bit errors. Two additional independent studies, on the other hand, one by Beckman Instruments, Inc., and the other by the U.S. Army Electronics Laboratories, concluded that errors occur in clusters or bursts, separated by relatively quiet periods of error-free transmission. The Alexander, Gryb, and Nast tests would probably have shown results similar to the other two studies if dropouts had been considered, which gives even more weight to the burst findings.

The bursts of errors most often seem to be caused by impulse noise (hits), dropouts (signal disappearance), and fades (loss of signal strength), which are circuit characteristics. The impulse noise tends to cause a conversion of low-frequency bits (usually marks) to high frequency bits (usually spaces), so is called *space biased.* Dropouts and fades tend to be mark biased.

Statements concerning error rates can be quite misleading if the *pattern* of errors is not given. If a circuit has a history of one error for every $10^6$ valid bits sent, this could mean that an error occurs after every millionth bit or so. But it also could mean that the circuit has never introduced an error, except once last year when the line was shorted for a few seconds, causing many errors to be generated.

### The Method - Slower Transmission

It is not difficult to understand that a slower transmission has more probability of being received correctly than does a faster one. Each segment (bit or character) of the message exists for a longer period of time, thus permitting detection devices a greater degree of assurance as to the presence of the signal. Noise, in other words, is a proportionately smaller part of the signal.

### The Method - Parallel Transmission

Parallel transmission probably is less subject to error than is serial, because in most cases it is slower. The present upper limit of parallel speeds over telephone lines is generally accepted as being about 75 cps, whereas serial transmissions are commonly 240 cps and above.

Parallel signals are held to lower speeds because speed is governed directly by bandwidth and the original channel must be divided into as many subchannels as there are bits in each coded character. The subchannels also must be separated from one another by a buffer zone of some adequate bandwidth, which further reduces the width of each subchannel. In addition, the subchannels often must be constructed to carry either of two frequencies for each bit of the code: one to indicate a mark, the other a space.

The time required to send a character serially is the cumulative time that it takes to send each bit of the character. Parallel transmission time, however, depends on the time required to send only one bit, since all bits comprising the character are sent simultaneously. Thus, if each bit of a 10-bit character sent serially lasted 1 ms, the character would be sent in 10 ms. If the same character were sent in 10 ms on a parallel system, each of its bits would also last 10 ms. This situation allows 10 times as long a period for the detecting device to determine the nature of the signal and greatly reduces the tendency of noise bursts to obliterate whole bits or groups of bits.

The diagram on the following page illustrates the resistance to error of parallel transmission (Fig. 5-3). The dark areas illustrate the presence of a noise burst on the two types of transmission. In the serial system, a noise burst of slightly more than 1 ms duration created a mark where a space existed. The same noise burst had no effect on the longer parallel character, since the receiver does not recognize such short signals. Parallel transmission also helps overcome delay distortion (and may be necessary in some cases), but is more sensitive than serial transmission to channel frequency displacement.

## ERROR DETECTION

### Use of Redundancy

In the previous example of the teletypewriter which received a signal to ring its bell, even after the entire five bits that made up the baudot bell code were received, there really was no way to tell with certainty that the bell code was the code that had been sent — one or more of the bits that comprised the code could have been changed during transmission.

A simple method of checking the validity of the transmitted codes would be to send each one twice. If one character of a pair differed from its mate, an error would have occurred. Or the entire message may be transmitted twice and every character compared. Errors would be found easily, for it would be improbable that identical errors would appear in identical places in both transmissions.

The obvious disadvantages to this type of error detection are its high redundancy (100 percent) and the lack of assurance as to which transmission was correct, the first or the last? As a result, attempts were made to both reduce redundancy and increase accuracy. Accuracy can be improved, but only at the cost of redundancy, so these two factors are used to test the worth of EDC codes.

### M-out-of-N (Fixed-weight) Codes

The baudot code has a minimum of redundancy. If only one bit in any character is reversed, the new configuration represents a valid character. To

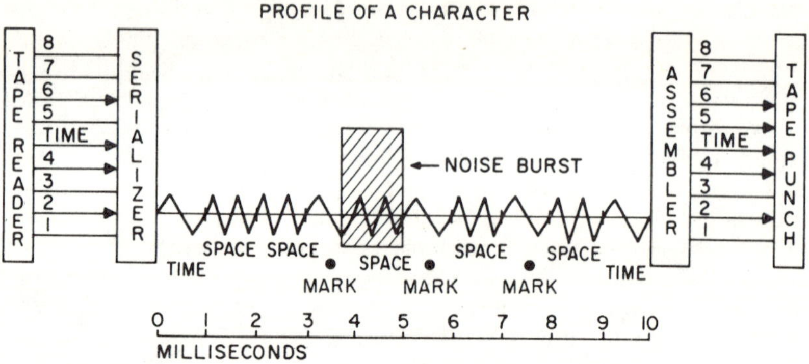
PROFILE OF A CHARACTER
TAPE READER
8
7
6
5
TIME
4
3
2
1
SERIALIZER
NOISE BURST
SPACE SPACE
TIME
MARK
SPACE
MARK
SPACE
MARK
SPACE
TIME
ASSEMBLER
TAPE PUNCH
0 1 2 3 4 5 6 7 8 9 10
MILLISECONDS

(A)

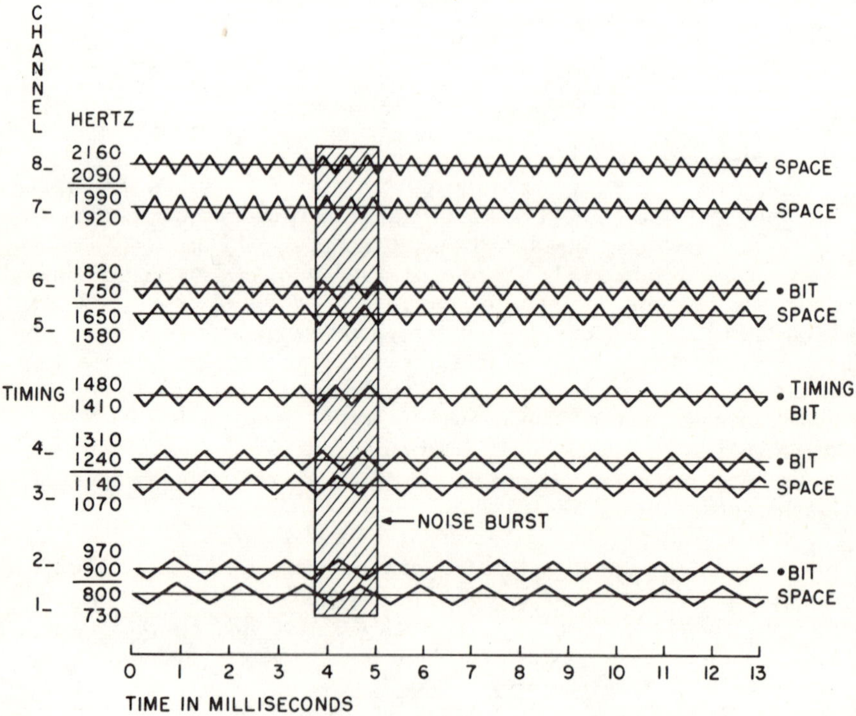
CHANNEL
HERTZ
8
2160
2090
7
1990
1920
6
1820
1750
5
1650
1580
TIMING
1480
1410
4
1310
1240
3
1140
1070
2
970
900
1
800
730
SPACE
SPACE
• BIT
SPACE
• TIMING BIT
• BIT
SPACE
NOISE BURST
• BIT
SPACE
0 1 2 3 4 5 6 7 8 9 10 11 12 13
TIME IN MILLISECONDS

(B)

be effective as an error-detection device, then, a code must be able to indicate in some way that one or more of its bits has been altered. One method of providing this capability is to arrange each character of a code so that it always has a constant ratio of one bits to zero bits.

The 2-out-of-5 code described earlier is one such code. There are always two 1's and three 0's representing a character; if this ratio is not present in the received character, an error has occurred. This system is better than none, but it has two drawbacks. First, although an odd number of bit errors in any one character will always be detected, an even number may not be caught. Second, the number of combinations, Q, that can be derived from a one-character group of bits is severely limited. A five-bit character such as that in the baudot code can have a maximum of $2^5$ or 32 discrete combinations. The 2-out-of-5 code, however, can have only two 1's and three 0's, and thus is limited to the results of the following equation, which is of the form M-out-of-N:

$$Q = \frac{N!}{M!\,(N-M)\,!}$$

Figure 5-3. *Effects of noise on serial and parallel signals. (A) A typical profile of a coded character being transmitted over a communications line. The code is first read and stored in a serializing buffer. Then it is transmitted as a series of impulses. A high frequency is used for the blank and a low frequency for the bit. (Note the lines on the chart indicating frequency show approximately 2 millihertz for a blank or 1 millihertz for a bit.) The equipment at the receiving terminal recognizes only the bit frequency. At the receiving terminal the bits are assembled in a buffer and fed to the tape punch. This is basically the system invented by Baudot about 1870 and until recently, employed in practically all equipment used to transmit data.*

*(B) A new concept perfected by the Bell Laboratories. In the new parallel system, each bit position in a tape code is assigned two frequencies, one to represent a bit and one to represent a blank. In this way, all eight levels plus the timing channel are transmitted simultaneously. The advantages of the system are (1) the serializing and reassembly buffers are eliminated, thus reducing manufacturing costs as well as possible points of failure; and (2) parallel transmission permits time to transmit each bit for a number of hertz, thereby greatly reducing the susceptibility to extraneous noise, a more or less constant hazard on the telephone line. The dark areas on the two diagrams illustrate the effects of similar noise bursts on the two types of transmission. In the serial system, a noise burst of slightly more than 1-ms duration completely blanks out one bit length and creates a bit where a blank actually was transmitted. In the approximately 13-ms pulse length of the parallel system, the same noise burst has no effect, since the receiver does not recognize such a short pulse. The new Data-Phone set using this principle has been assigned the numbers 402A (transmit) and 402B (receive) and is currently in production. (Courtesy of Tally Corp.)*

Substituting:

$$Q = \frac{5!}{2!(5-2)!}$$
$$= \frac{5 \times 4 \times 3 \times 2 \times 1}{2 \times 1 (3 \times 2 \times 1)}$$
$$= 120/12 = 10 \text{ legal combinations}$$

which limits this code to expressing only numerical data (unless augmented by zone bits).

The popular 4-out-of-8 code has both limited combinations and a comparatively high redundancy, making it undesirable for communication use:

$$\text{Combinations, } Q = \frac{N!}{M!(N-M)!}$$
$$= \frac{8!}{4!\,(8-4)!}$$
$$= \frac{40{,}320}{24 \times 24} = \frac{40{,}320}{576} = 70$$

The information content of each character of this code is found by taking the logarithm to the base two of the number of possible characters:

$$I = \log_2 70 = 6.13 \text{ bits}$$

Redundancy of the 4-out-of-8 code, then, is:

$$R = 1 - (\text{Bits}_I/\text{Bits}_T)$$
$$R = 1 - (6.13/8) = 1 - 0.767 = 0.233 \text{ or } 23.3\%$$

One advantage often stressed concerning the 4-out-of-8 code is its inherent error-detection capabilities, but a comparable seven-bit code using an eighth bit for parity would have $2^7 = 128$ combinations and a redundancy of only 12.5 percent.

### Modulus-N Check Digits

Another special-purpose but very useful form of redundancy is the modulus-N check digit. Assume that payroll numbers, vendor numbers, or some other recurring series of numbers are to be transmitted, and there is a requirement for a high degree of accuracy in the received data. It would be desirable to have some way to check these figures after receipt, particularly if they are to be entered into the system manually.

A typical number of this type might be 27143. To this we can append a sixth digit, X, for check purposes. To be of any use, the digit X must have a particular relationship to the original number. Digit X may be generated by multiplying each digit of the original number by a series of weights, then adding the products:

$$\begin{array}{ccccc} 2 & 7 & 1 & 4 & 3 \\ \times & \times & \times & \times & \times \\ 1 & 3 & 1 & 3 & 1 \\ \hline 2 & +21 & +1 & +12 & +3 = 39 \end{array}$$

In the modulus-10 system, the sum would be subtracted from the next higher multiple of 10 and the result would be digit X: 40 − 39 = 1, which would be appended to the original number to form a modulus check number, 271431. A device at the receiving end performs the same operation on the number and should arrive at the same modulus check digit. If the check fails, an error has occurred.

The system is called modulus-N because any number can be substituted for N; 10, 11, and 2 are the most used (2 for binary systems). The weights also can have arbitrary sequences, such as 1-2-3-4 . . . , 1-3-1-3-1 . . . , . . . 4-3-2-1, and so on. Several data origination devices have the capability of generating modulus-N check numbers as an aid toward error control and of punching them into paper tape for transmission.

### Odd or Even Parity

A simple device to detect bit rearrangements which occurred during transmission or processing is to add a parity bit to the bits which already describe each character. The addition of parity 1 bits to only those characters with an odd number of 1 bits of their own will produce a code in which all characters have an even number of 1 bits. Thus a reversal of the value of any *one* bit will produce an invalid (odd) combination. This method of odd or even parity is a special case of the M-out-of-N method.

### Horizontal and Vertical Parity

When one parity bit is added to each group of bits describing a *character* as above, it is called *vertical* parity. This comes from the relative position of the bits if they are observed on a paper tape held lengthwise. *Horizontal* parity, as might be guessed, results in an extra bit being placed at the end of a block of bits along the *length* of the tape. If horizontal parity is supplied for each channel in the tape, a *parity character* will be created following a block of information characters (see Fig. 5-4).

The use of a combined horizontal plus vertical parity overcomes one of the greatest weaknesses of either scheme alone: The inability to locate the bit(s) in error exactly. In the array shown in Fig. 5-5, the receiving device has checked the horizontal and vertical parity of the incoming characters and tagged those that indicated an error by wrong parity. By reversing the values of the error bit, the receiving device can correct the character.

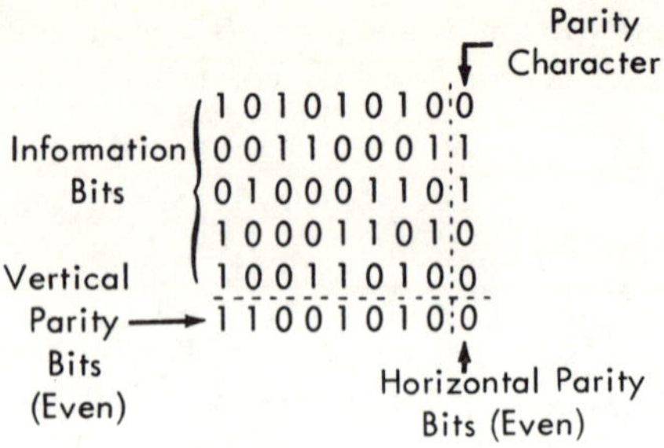


Figure 5-4. *Vertical and horizontal parity.*

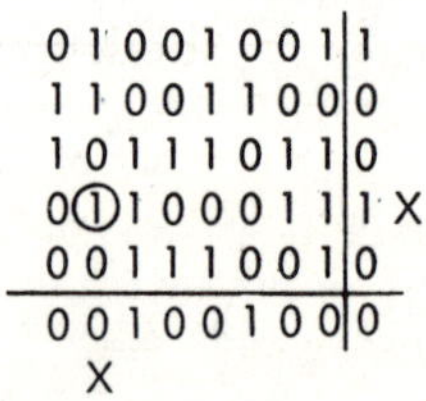


Figure 5-5. *Combined horizontal and vertical parity.*

Even though a combined horizontal and vertical parity scheme offers much more protection than either one alone, it has one drawback: any rectangular configuration of errors will not be detected. Fig. 5-6 shows a possible (but highly improbable) situation in which the errors would not be detected by combined horizontal and vertical parity. If the bits within the circles were reversed in value by some error-causing disturbance, the parity bits would remain valid and the error would go undetected.

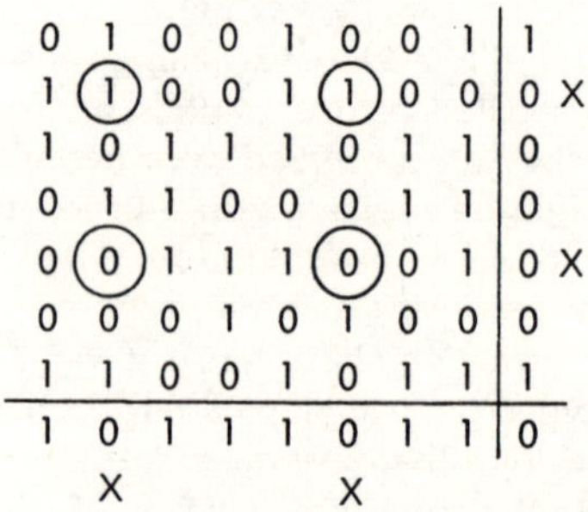


Figure 5-6. *Errors not detected by combined horizontal and vertical parity.*

**Interlaced Parity**

To overcome this weakness of horizontal and vertical parity to some extent, another method of error detection called interlaced parity can be used. This method requires two extra parity bits per character; the first parity bit provides even or odd parity for the first, third, and fifth bits of a six-bit code, and the second parity bit would do the same for the second, fourth, and sixth bits. Fig. 5-7 indicates how even interlaced parity woud be generated for a six-bit coded character. Interlaced parity can greatly reduce the probability of an undetected error as compared to an unchecked system. This method is also called a *cyclical redundancy check.*

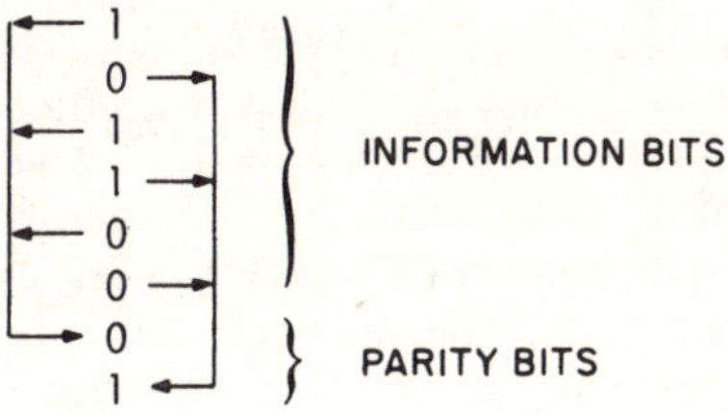


Figure 5-7. *Generation of even interfaced parity.*

Any parity error-control system has one weakness: the possibility that the parity itself may be received in error. Unless the data error corresponds exactly, however, the error will be detected, and can be corrected by retransmission.

## ERROR CORRECTION

**Retransmission**

Although more sophisticated methods will be discussed below, the simplest, most effective, and least expensive method to correct errors usually is to detect them by using a reliable detection system and causing the erroneous data to be retransmitted. Correction by retransmission, used by most data equipment, requires comparatively simple logic and little storage, whereas other methods require more expensive terminal equipment and elevated transmission costs.

The most popular error detection and correction devices on the market use retransmission as a correction device. They transmit blocks of characters and make appropriate checks for valid parity. If the parity generated at the transmitter checks with that which has been created from the received message by the receiver, the transmission continues. If the parity check fails, the last block is retransmitted and checked again for parity. This method avoids the disadvantages of transmitting the entire message twice and of having to compare the second message with the first for validity.

## Hamming Distance

The number of bits by which character codes differ is called the hamming distance. For example, the baudot code contains characters that differ from one another by the placement of only one bit. Changing one bit of the letter A could give any of quite a variety of letters. If a parity bit were added to each character, each character would then differ from all other characters by two bits, and a one-bit error would not create another valid character.

A hamming distance of one between characters, such as in the baudot code, will not permit even error detection because all combinations are valid. The addition of a parity bit to each character gives a hamming distance of two, which permits error detection, but not correction. This is because a one-bit error within such a character will be recognized, but there is no provision to identify the particular error bit. The one-bit error does not cause the character to resemble any other character, however. Two bits in error, though, will cause the parity to again be valid and could cause a complete change in the meaning of the character, as shown in Fig. 5-8.

| | | | |
|---|---|---|---|
| BITS IN ERROR: | NONE | ONE | TWO |
| CHARACTER: | A | (NONE) | Z |
| CHARACTER CODE | 1 | 1 | 1 |
| | 1 | 0 | 0 |
| | 0 | 0 | 0 |
| | 0 | 0 | 0 |
| | 0 | 0 | 1 |
| EVEN PARITY: | 0 | 0 | 0 |
| ERROR DETECTED: | — | YES | NO |

Figure 5-8. *Errors not detected by parity.*

The detection *and correction* of one error is permitted by a hamming distance of three (assuming the redundancy is properly placed), since a character with a one-bit error more closely resembles its original meaning than the meaning of any other character in the set. A hamming distance of four will permit the detection of two errors or the correction of a single error.

This concept may be clarified by using geometrical models. In these examples, a three-digit code will be used for clarity; the same principles hold for codes of any length.

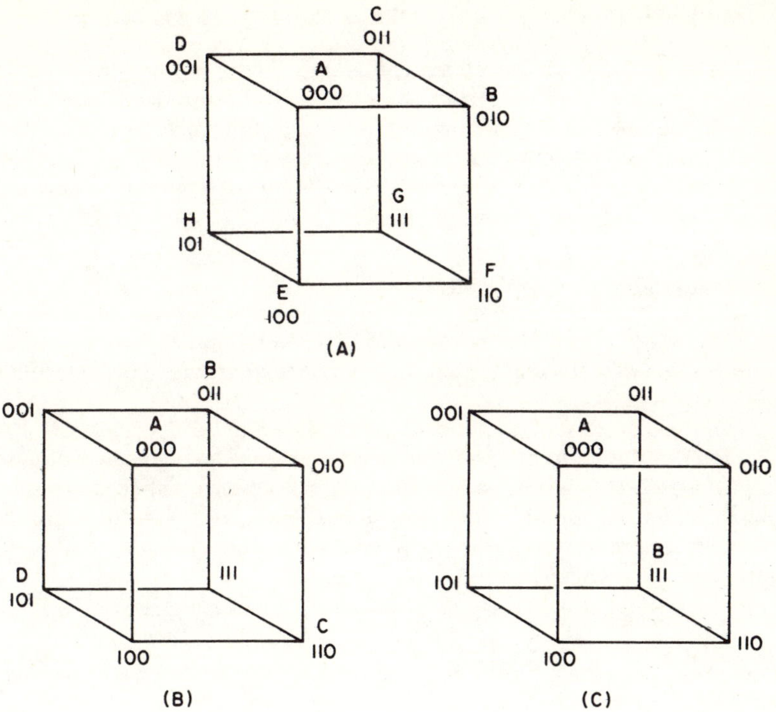


Figure 5-9. *Geometric models.*

The cubes shown in Fig. 5-9 represent mathematical models of a code structure. A unique code is located at each of the eight vertices of the cubes, and each code differs from any adjacent code by one digit. If a letter is assigned to each code group (Fig. 5-9(A)), the hamming distance (number of cube edges) between valid codes is one, and the code has no error detecting nor correcting capabilities. This is true because any code received resembles a valid letter combination, and there is no way to tell if the code structure was altered during transmission.

If the number of valid codes is reduced to four (Fig. 5-9(B)), an invalid (error) code is obtained between each valid (letter) code. The hamming distance is increased to two and the property of error detection is introduced; if any one bit of a valid code changes value, it becomes an error code. Upon receiving an error code, however, there is no way to tell what it should have been, and so the capability of error correction is lacking.

Figure 5-9 (C) has only two valid codes and six error codes. If any one bit changes value in a valid code, it will still more closely resemble its original value than the value of the other valid code, and the bit in error can be corrected. This single-error correcting code has a hamming distance of three, and a redundancy of 67 percent.

## Forward-acting Codes

As mentioned above, codes containing simple parity permit the detection of errors, but no means of correction. If redundancy is added to original messages in sufficient quantities and in the correct patterns, codes can be constructed so that messages received with errors can be corrected at the receiving end without retransmitting the message or any portion of it. Several such codes are discussed below.

## The Matrix Sum

The Navy uses radio teletypewriter (RATT) circuits extensively and requires that data transmitted on these circuits arrive with a predetermined degree of accuracy.[1]

In the matrix-sum method of error detection and correction, each character of the set to be used is arbitrarily assigned an equivalent number: A = 1, B = 2, C = 3, and so on. The original message is then divided into equal groups of two or more characters each, and the corresponding numbers are substituted for each character. For example, if the data consisted of the characters

A B C D E F G H I J K L M N O P Q R

they could be grouped in threes, and their respective numbers assigned as follows:

| ABC | DEF | GHI | JKL | MNO | PQR |
|---|---|---|---|---|---|
| 123 | 456 | 789 | 10 11 12 | 13 14 15 | 16 17 18 |

Now the numbers can be arranged in a square matrix

1 2 3

4 5 6

7 8 9

and the vertical and horizontal sums taken

$$
\begin{array}{rcrcrcr}
1 & + & 2 & + & 3 & = & 6 \\
4 & + & 5 & + & 6 & = & 15 \\
7 & + & 8 & + & 9 & = & 24 \\
\hline
12 & + & 15 & + & 18 & = & 45
\end{array}
$$

In all cases, the total of the vertical sums equals the total of all the horizontal sums and becomes the check digit. If each of these sums is then

[1]The discussion of the matrix-sum method is taken from an article by LCDR E. J. Hauber in *Naval Communications Bulletin,* February, 1965. The steps described are accomplished by computers at both the transmitting and receiving locations.

converted back into an equivalent letter, where A = 1, B = 2 . . . , Z = 26, A = 27, B = 28, C = 29, . . . , and the like, the matrix would appear as

| | | | |
|---|---|---|---|
| A | B | C | F |
| D | E | F | O |
| G | H | I | X |
| L | O | R | S |

For transmission, the message would be rearranged to:

A B C D E F G H I L O R F O X S

At the receiving station, the first nine letters are again arranged into groups and the above process repeated. Any deviation from the previous results indicates an error. This system will permit detection and correction of any garble consisting of three or less consecutive errors. For example, assume that the above message is received as:

A B C X Y Z G H I L O R F O X S

The matrix of corresponding numbers and sums would be

| | | | |
|---|---|---|---|
| 1 | 2 | 3 | 6 |
| 24 | 25 | 26 | 75 |
| 7 | 8 | 9 | 24 |
| 32 | 35 | 38 | 105 |

and the letters corresponding to the sums would be

| | | | |
|---|---|---|---|
| | | | F |
| | | | W |
| | | | X |
| F | I | L | A |

With the exception of the F and X in the vertical column, these do not agree with the letters transmitted.

To determine where the errors exist, the first step is to convert the seven check letters to their equivalent numbers and see if their sums agree:

| | | | |
|---|---|---|---|
| | | | 6 |
| | | | 15 |
| | | | 24 |
| 12 | 15 | 18 | 45 |

The sums agree, so the check letters were received correctly.

The next step is to fill in the rest of the equivalent numbers and add both vertically and horizontally:

| | | | |
|---|---|---|---|
| 1 | 2 | 3 | 6 |
| 24 | 25 | 26 | 15 |
| 7 | 8 | 9 | 24 |
| 12 | 15 | 18 | 45 |

It is apparent that the middle horizontal row is incorrect. It is only a matter of simple subtraction in the vertical columns to determine the numbers which should appear in the second row. The same procedure may be used no matter where the garble appears.

While it is apparent that this process is tedious, it is well suited for computer applications. It is particularly important to note that this form of error correction does not use any feedback circuit, and thus can be used in one-way transmissions.

The three-level matrix used above will correct about 90% of all errors. For more accuracy, matrices of more levels can be used. A nine-level matrix, for example, will correct 99.9 percent of all garbles, which gives an undetected error rate of one bit to 300,000. The nine-level matrix also makes more efficient use of circuit time with an increase of 23.4 percent in message length as compared to the 78 percent additional length imposed by the three-level system.

**Hamming Code**

Another EDC code is the hamming code, after R. W. Hamming, who did early work in error correction. The code introduces a hamming distance sufficient to accomplish error detection, correction, or both. This is done by associating even-parity check bits with unique combinations of information bits.

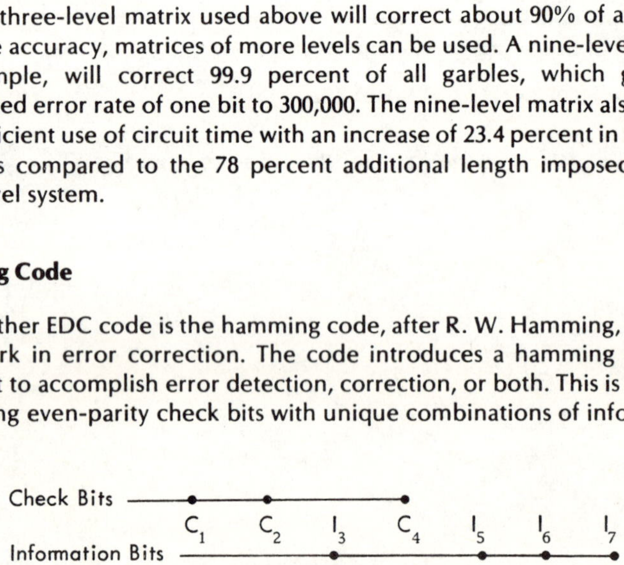


Figure 5-10. *The hamming code.*

Using a four-bit code as an example, a character might be represented by 0101, with no parity bit. To add parity by the hamming method, three parity check bits $C_1$, $C_2$, and $C_4$, are inserted, resulting in the seven bit code shown in Fig. 5-10.

Each check bit is associated with its own particular combination of information bits; $C_1$ provides even parity for information bits $I_3$, $I_5$, and $I_7$; $C_2$ does the same for $I_3$, $I_6$, and $I_7$; and $I_5$, $I_6$, and $I_7$ are checked by $C_4$, as shown in Fig. 5-11.

Now, substituting the example character, 0101, for the information bit positions:

$C_1\ C_2\ O\ C_4\ 1\ 0\ 1$

and by having each of the check bits give even parity to the information bits they monitor, the following hamming code is equivalent of the original 0101:

0 1 0 0 1 0 1

Assume that during the transmission of this hamming character, $I_5$ is altered from a 1 to a 0 by an error condition. Since this bit is monitored by $C_1$ and $C_4$, both $C_1$ and $C_4$ will now show odd parity for their bit combinations. $C_2$, which does not monitor $I_5$, will have retained its original

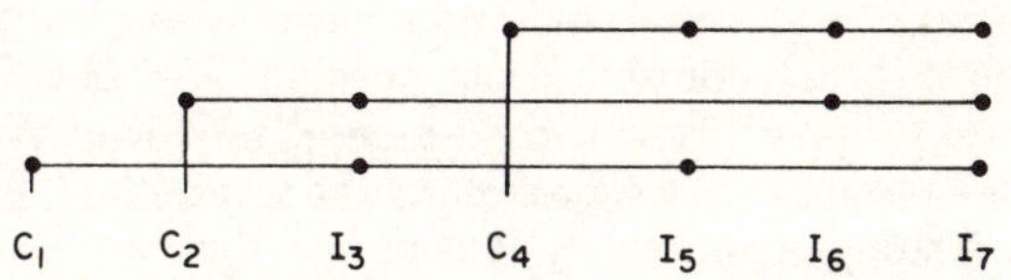


Figure 5-11. *Relationship of hamming check bits to information bits.*

even parity. By gating the check pulses into appropriate registers, the receiving device can determine the validity of the received character, correct the error bit, strip it of the check bits, and deliver the original configuration (0101). In fact, if the check bits are arranged properly, as in Table 5-3, the check bits which have odd parity will form a binary number which pinpoints the position of the bit in error. As seen in the example above, if $I_5$ was received in error, $C_1$ and $C_4$ would show odd (error) parity. As the table shows, if $C_1$ and $C_4$ had odd parity with the bits they monitored then $I_5$ was the bit in error.

Important — a 1 in the table means that the check bit is indicating a parity error. It is not necessarily the actual value of the check bit. Notice, too, that the check bits form a binary number which, when translated to decimal, indicates the numerical location of the error bit (101 = 5 = $I_5$).

**TABLE 5-3. BINARY PROPERTY OF HAMMING CHECK DIGITS**

| Check Bits Showing Parity Error $C_4$ | $C_2$ | $C_1$ | Error Bits |
|---|---|---|---|
| 0 | 0 | 0 | None |
| 0 | 0 | 1 | $C_1$ |
| 0 | 1 | 0 | $C_2$ |
| 0 | 1 | 1 | $I_3$ |
| 1 | 0 | 0 | $C_4$ |
| 1 | 0 | 1 | $I_5$ |
| 1 | 1 | 0 | $I_6$ |
| 1 | 1 | 1 | $I_7$ |

Its hamming distance of three enables the code to detect and correct one error. The error is corrected by simply reversing the value of the bit that is incorrect.

## Hagelbarger Code

While the hamming code will correct only one error within a block of some assigned size, the hagelbarger code will correct up to six consecutive errors without the restriction of blocking the transmitted data. The hagelbarger code does require, however, that the error group be followed by at least 19 valid bits before further error bits are encountered.

Figure 5-12 shows the hagelbarger encoder. Message bits enter the seven-bit register one at a time. After each new bit enters the register, the contents of the first and fourth register positions are tested, and an even parity (modulus-2 sum) of these two positions is produced by the parity bit generator. The output of the last position of the seven-bit register and of the parity bit generator are alternately sent out on the transmission line (100 percent redundancy). Thus each message bit is associated, by parity, with the fourth message bit both preceding and following it.

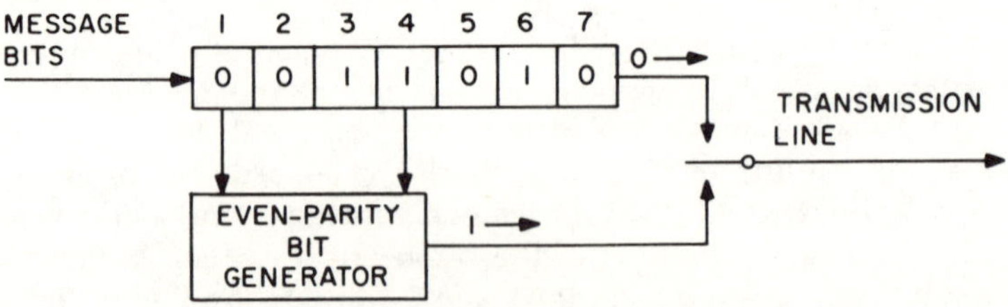


Figure 5-12. *Hagelbarger encoder.*

The hagelbarger decoder is described by Fig. 5-13. As the bit-stream is received, the message bits are directed to the first position of the message register, while the parity bits are shunted to a shift register. Two parity checks are made after each parity bit (and its accompanying message bit) are received. The A parity check tests positions 1 and 4 of the message register, while B tests positions 4 and 7. If both tests fail (are odd), the bit in the fourth position is in error and is reversed before it is sent on to position 5. If the A check shows even parity and the B shows odd, the tenth position of the shift register contains an error and should be reversed before being used.

## Bose-Chaudhuri Code

Another popular forward-acting code, devised by R.C. Bose and D. K. Ray-Chaudhuri, has a hamming distance of 5 and is thus capable of correcting double errors. Its popularity stems from its power to *detect* up to four errors.

Both Beckman Instruments and Bell Telephone have considered using the error detection properties of a bose-chaudhuri code in their equipment, and both have elected to correct by retransmission rather than use the less efficient *error-correcting* properties of the code. The code was originally designed to introduce 10 check bits for every 21 bits of data. The Bell version would use 12 check bits for 48 data bits, while the Beckman device would use 30 check bits with every 992 data bits.

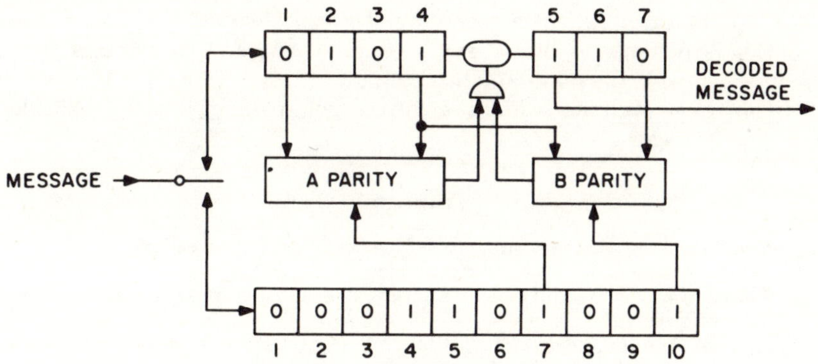


Figure 5-13. *Hagelbarger decoder.*

## CHOOSING AN ERROR-CONTROL SYSTEM

Error-control systems come in as many varieties as do data processing systems. Careful consideration must be given to all facets of the system to be sure of a maximum control of errors at a minimum loss of throughput. Listed below are a few factors that should be closely evaluated during the early stages of system design.

(1) What are the extent and pattern of error-inducing conditions on the circuits to be used?
(2) What is the maximum error rate that can be tolerated?
(3) What are the effects of no error control? Detection alone? Detection and retransmission? Detection and correction?
(4) How does the cost of increased accuracy compare with present costs (apparent and hidden) of correcting errors and error-caused conditions?
(5) What are the length, frequency, urgency, and content of present or future messages?
(6) What degree of redundancy do various systems require to achieve the desired degree of accuracy?
(7) How do different systems compare as to over-all transmission efficiency?

(8) What block lengths provide the highest efficiency for various error conditions?
(9) What portion of the error-control system provides the required parity? Block characters?
(10) Should errors remain in the received medium? How should they be flagged?
(11) Will the introduction of error control require any computer reprogramming?
(12) What are the tradeoffs between the relatively large amount of turnaround time caused by short data blocks, as compared to increased retransmission time of longer blocks?
(13) What mix of block length, redundancy, and turnaround time provide the maximum throughput?
(14) How well do various error-control methods match available storage?

## Problems

1. Give four things that may affect the efficiency of a communication system.
2. What is redundancy? Give some examples.
3. In what ways might redundancy be undesirable? What purpose does redundancy most often serve?
4. Calculate the redundancy of a six-bit code using a seventh bit for parity. Show work.
5. What is the efficiency of the code described in Problem 4? Calculate in two ways, and show work.
6. If a code has an efficiency of 87%, a transmission system with an efficiency of 67%, and transmission speed rated at 2400 bps, what is the effective information transfer rate in bps? Show work.
7. What factors may affect the effective speed of a system?
8. What is a code element, and of what use can it be?
9. Describe several methods of reducing the number of errors in received messages.
10. Give two disadvantages of an "M-out-of-N" code.
11. How many characters can be described uniquely by a 5-out-of-8 code?
12. What is the information content, in bits, of each character of the code given in Problem 11?
13. What is a "forward-acting" code? What would you say about the relative redundancy of any such code? Why?
14. List six or more factors that should be carefully considered in evaluating an error-control system.

*CHAPTER 6*

# *Modulation and Multiplexing*

## MODULATION DEFINED

Modulation may be defined as the modification of some characteristic of an otherwise continuous signal called a *carrier*. Usually, a carrier is a single-frequency electronic wave which serves as a means of propagation. It is modified, in some way that can be accurately detected at the receiving end, by a signal which contains the information we wish to transmit. This signal is usually a "baseband" signal, which will be discussed shortly.

Characteristics of the carrier that may be modified by the modulating signal include its amplitude, frequency, phase, or a combination of these. The complexity of the resultant modulated signal depends on (1) the kind of modulation being performed (the characteristic(s) being changed) and (2) the number of discrete states or conditions the signal must represent. The modulation of binary data, for example, requires only two discrete states of the amplitude, frequency, or phase of the modulated carrier. These two states can readily be distinguished by the receiving demodulator and reconverted to a likeness of the original signal. As the modulating signal becomes more complex, that is, assumes three, four, or more states or levels, the modulated carrier must also assume a greater number of states, and the receiving equipment must be more sophisticated to detect the less obvious differences in signal conditions.

### Baseband Signals

The construction and composition of various communication and computer codes were discussed in a previous chapter. A business machine, as a result of converting a coded character to a signal to be transmitted over communication facilities, produces a series of d-c pulses and no pulses



ABERYSTWYTH LIBRARY

(binary ones and zeros). These are in sequences which uniquely define the character which they represent, and are shown graphically in Fig. 6-1(A) as they would appear on a circuit.

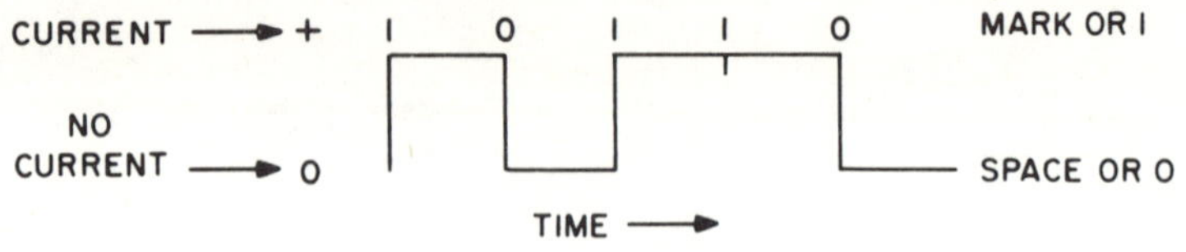


(A)

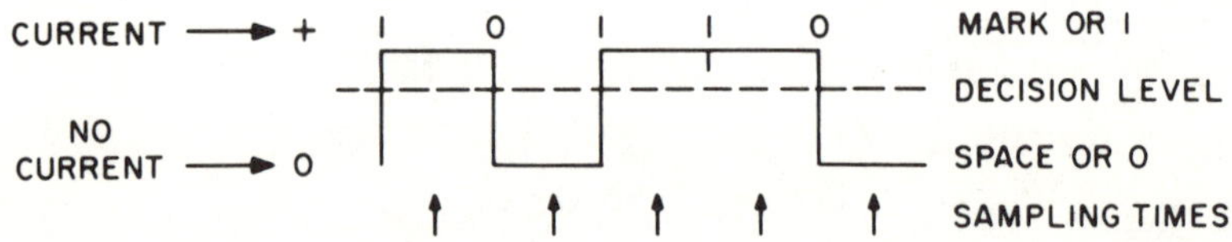


(B)

Figure 6-1. *(A) An example of a baseband signal; that is, the unmodulated d-c signal produced by an information source such as a business machine. (B) The signal reproduced.*

The series of bits (10110) in Fig. 6-1(A) represents the value of the following signal bit. A dotted line indicating the *decision level* has been drawn through the approximate middle of the signal's magnitude. The binary value of the signal at any instant is determined by its presence above or below the decision level (sometimes called the slicing level).

## Sampling

Sampling, a technique for determining the value of a signal, is done by the receiving business machine in one of several ways. One method causes a sampling of the signal near the midpoint of the duration of each pulse. Another method uses circuits that will change state each time the signal level advances or retreats across the decision level. If the properties of the baseband signal are such that, rather than changing from a zero value to a positive value, it changes from a *negative* to a positive value, then another method of sampling involving the detection of the zero-crossing of the signal may be used. In summary, the purpose of sampling is to determine the value (1 or 0) of a portion (bit) of a received signal.

**Examples of Basebands**

Various types of baseband signals have been developed to meet special needs; several of the most common types are shown in Fig. 6-2. The same bit pattern is used for each illustration, and the signals are positioned vertically to dramatize their differences.

The choice between baseband signals for use in a particular system may be arbitrary. Each type of signal has characteristics, however, that may make it more or less suitable under given circumstances. The *return to zero-space* is the simplest baseband signal. It consists of a d-c current for a mark and no current for a space. This signal was used for early key telegraphy, and is now used with private-line teletypewriter systems. Sampling determines the presence or absence of current.

Rather than have current for a mark and no current for a space, it is often desirable to have a definite current for both. A polar baseband signal uses positive d-c current for a mark and negative current for a space. The decision level in this case is zero current. Since there is no transition between two consecutive bits of the same value, sampling is used to determine the value of each received bit.

Neither of the above two signals gives a definite transition after each transmitted bit. To overcome this, the mark pulse of the return-to-bias signal was shortened to allow the signal value to return to its negative bias after each mark current. There is no change following a negative space current.

The first two signals described above would not change state if succeeding bits had the same value. It is often desirable, for purposes of synchronization, for example, to have the signal change state for each bit transmitted. The return to zero signal produces a short negative current for a space, and has a short positive current for each mark. Following each bit transmission, the signal will return to its zero value until the next bit is sent. This signal does not require sampling, since there is a pulse of discrete value for every bit.

A variation of the return-to-zero signal is the bipolar signal, which gives a short pulse in alternating directions for successive marks with an immediate return to zero current. Spaces cause no change from the normal zero current. Since every bit does not cause a transition, synchronized sampling is required to determine the state of each bit.

The nonreturn-to-zero baseband signal responds with a transition to the opposite value of current for each mark. There is no return to a bias, and a space does not cause a transition. The state of each bit must be determined by sampling.

**Purposes of Modulation**

The output of most business machines is in the form of baseband signals such as those discussed above. These signals have a d-c component that

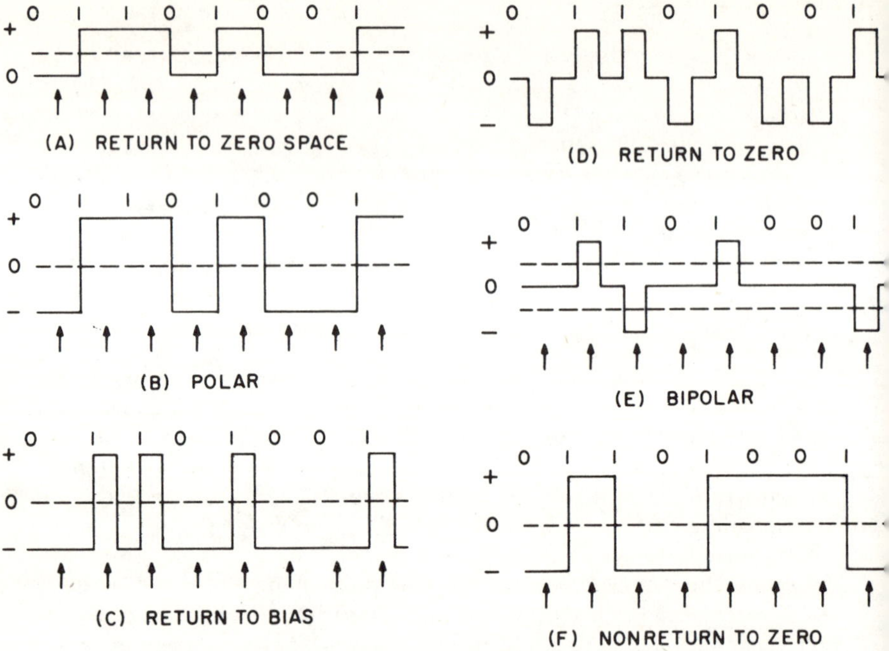


Fig. 6-2. *Examples of baseband signals: (A) Return to 0 space: simplest form of baseband signal. Consists of pulsating d-c current. Requires sampling to determine state of each bit. (B) Polar: positive current for 1-bits, negative for 0-bits. Requires sampling to determine state of each bit. (C) Return to bias: short positive current for 1-bits, no change from negative bias current for 0-bits. Requires sampling to determine state of each bit. (D) Return to 0; short positive current for 1-bits, negative for 0-bits. Return to 0 current after every bit. Requires no sampling because there is a pulse for every bit. (E) Bipolar: short pulse in alternating directions for successive 1-bits, with immediate return to 0 current. No change for 0-bits. Requires sampling to determine state of each bit. (F) Nonreturn to 0: pulse in alternating directions for successive 1-bits. No change from existing bias for 0-bits. Requires sampling to determine state of each bit.*

can't be transmitted through the switched message network, carrier systems, or radio systems. To be transmitted through most commercial communications channels, then, these baseband signals must be used to modulate a carrier of a higher frequency.

The baseband signal from a machine transmitting at a steady rate of N bits per second, for example, may have a frequency bandwidth ranging from 0 Hz while several 1 bits are sent in series (the d-c component), to N/2 Hz, while a series of alternate 1 and 0 bits are sent (the highest frequency). Somewhere between the frequencies of 0 and N/2 Hz, the transmission system will exhibit a cutoff, which blocks all frequencies from that point on down to 0 Hz. If this baseband signal is used to modulate a carrier within the voice band, however, the entire bandwidth from 0 to N/2 will be relocated to within the voice band, and the frequencies below the former cutoff point will not be lost (see Fig. 6-3).

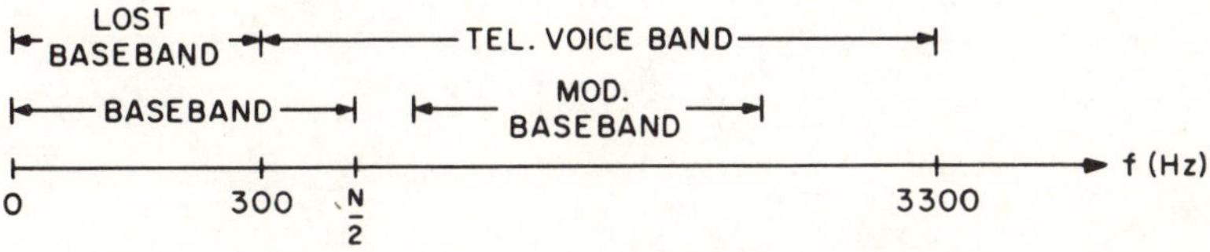


Figure 6-3. *Relocation of baseband signals by modulation.*

Another use for modulation is to relocate voice, data, and other signals to less-congested frequency bands. This results in the ability to use less noisy areas in the frequency spectrum, and, more importantly, it results in more efficient use of available bandwidths. A baseband teletypewriter signal, for example, has a bandwidth of up to 150 Hz, which occupies only a small portion of the 3300 Hz bandwidth provided by a voice channel. In turn, up to 24 voice channels may be carried by one pair of wires through use of modulation techniques. Yet, without modulation, the same pair of wires would be limited to carrying the d-c teletypewriter signal alone.

A third use for modulation is to provide more efficient radiation of energy through the use of higher frequencies; baseband signals comprise dc and possibly low-frequency ac, which are quickly absorbed by the internal resistance of a transmission system.

## TIME AND FREQUENCY DOMAINS

One of two references could be used to represent graphically a signal of frequency, *f*: time or frequency. They are interdependent, of course, but they are portrayed differently. Fig. 6-4 shows a frequency, *f*, as it would appear in both the time domain (A) and the frequency domain (B). The time domain shows amplitude on the ordinate and time on the abscissa. The frequency is indicated by the number of cycles within time *t*. The frequency domain also shows amplitude on the ordinate, but the frequency itself is indicated by the location of a spike on the abscissa.

For every change in the time domain, there is a corresponding change in the frequency domain. (See Fig. 6-5.) Notice that a greater number of cycles within time *t* in the time domain results in a shift to the right (higher position) of the frequency spike in the frequency domain:

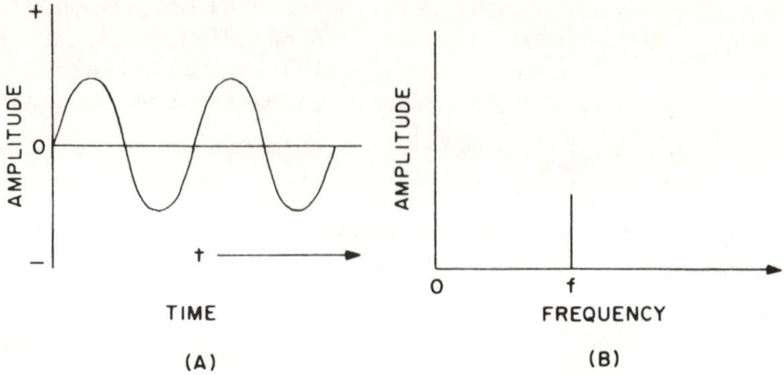


Figure 6-4. *Comparison of frequency, f, in (A) the time domain and (B) the frequency domain.*

The three-dimensional diagram in Fig. 6-6 combines the domains of time and frequency shown in the previous figures and thus may help to clarify their relationship. Notice how the waveform moves to the right along the frequency line as it increases in frequency. Single frequencies, then, may be represented by a single line on the frequency-domain diagrams (Fig. 6-7A), and bandwidths (or passbands) may be designated as being encompassed by two frequencies (Fig. 6-7B).

## TYPES OF MODULATION

### Continuous-wave (CW) Modulation

Messages sent in the familiar dits and dahs of the morse code are probably sent by continuous-wave (CW) modulation. In CW modulation, a

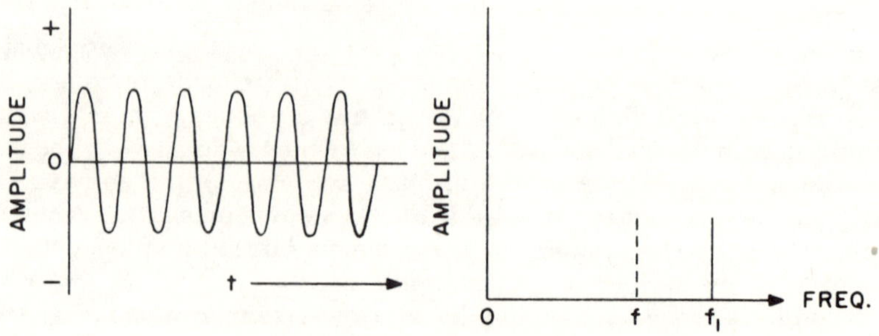


Figure 6-5. *Effects of higher frequency on time and frequency domains.*

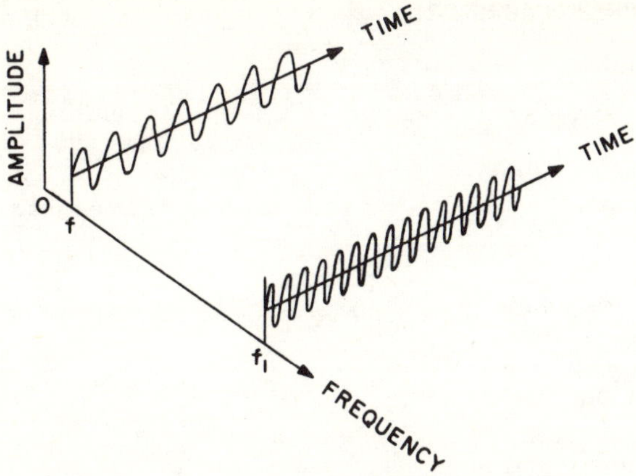


Figure 6-6. *Relationship of time and frequency domains.*

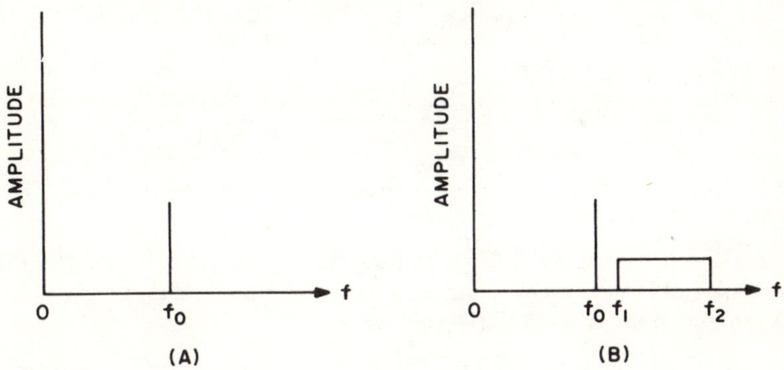


Figure 6-7. *Representation of (A) single frequencies and (B) bandwidth.*

carrier of some steady frequency (say, 1000 Hz) is generated by an oscillator. The carrier is transmitted when a key is depressed, and it stops when the key is released. The signal, then, is of the pattern in which the key is manipulated, and the carrier is simply turned on and off accordingly to this pattern (see Fig. 6-8).

## Frequency-shift Keyed (FSK) Modulation

A form of frequency modulation not much more complicated than CW is frequency-shift keying (FSK). Here, two frequencies are used: one is transmitted when a key is depressed, the second when the key is released. Such a method is used extensively with present teletypewriter transmission

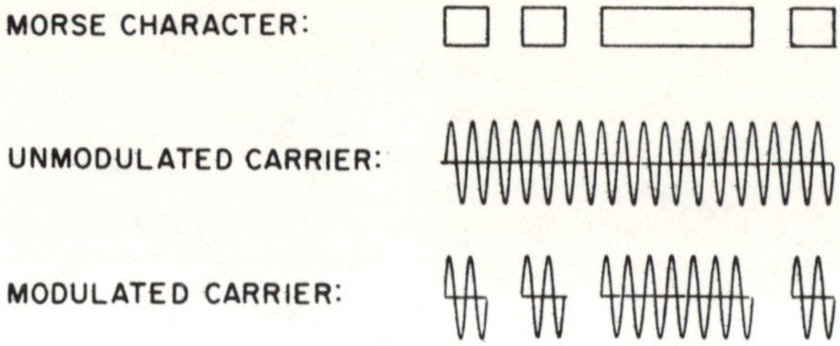


Figure 6-8. *Continuous-wave (CW) modulation.*

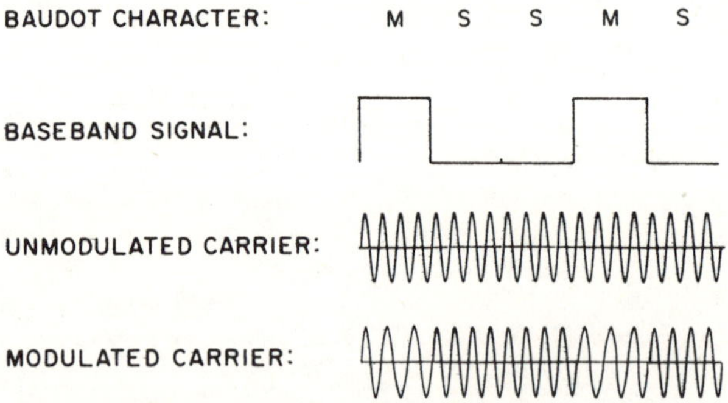


Figure 6-9. *Frequency-shift keyed (FSK) modulation.*

systems; one frequency is used as a mark, another as a space. The main advantage of FSK over CW is that steady power is applied to the transmitter, maintaining peak efficiency. (See Fig. 6-9.)

## Amplitude Modulation (AM)

Both CW and FSK employ frequencies of constant amplitude. Amplitude modulation (AM) differs in that the amplitude of a steady carrier frequency, C, is varied in relation to a modulating signal frequency, M, that is impressed upon it. The result is a single-frequency carrier within an "envelope" of varying amplitude, as shown in Fig. 6-10.

To produce this envelope effect, a low-frequency modulating wave, M, is superimposed on a higher-frequency carrier wave, C, producing two product frequencies, C + M and C − M, called *sidebands.* Consider the following example. Fig. 6-11 shows a frequency range beginning at 0 Hertz and extending upward. The acoustical range extends roughly from 20 Hz to 20,000 Hz. A data frequency, V, is shown at some point within the bandwidth of the switched telephone network, between 300 and 3300 Hz. At some point at a much higher frequency a carrier, C, is shown. The data frequency, V, has been applied to the carrier, thus modulating it, and the results of this

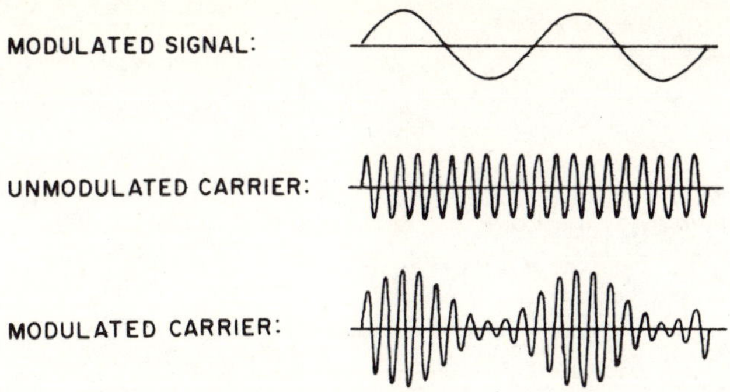


Figure 6-10. *Amplitude modulation (AM).*

modulation are the sideband frequencies C − V (lower sideband) and C + V (upper sideband). These sideband frequencies have an amplitude of V/2 and are at a higher position in the frequency spectrum than is V, but V and both sidebands all carry the same intelligence.

If C had a frequency of 100,000 Hz, and V a frequency of 1000 Hz, then C + V would be 101,000 Hz, and C − V would be 99,000 Hz. It is not common, though, to have a modulating wave of a single frequency as in the example above. Usually the carrier will be modulated by a frequency that varies

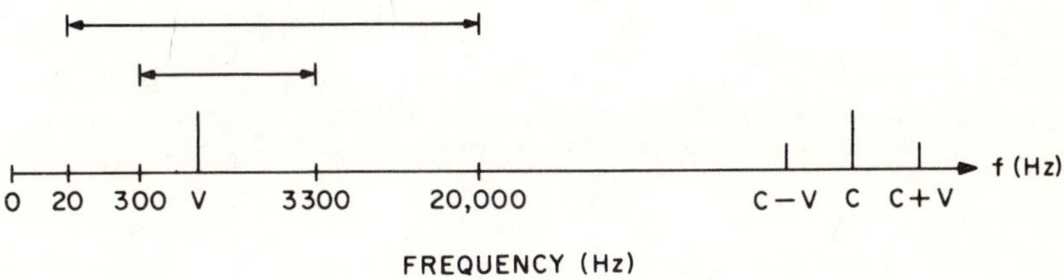


Figure 6-11. *The relocation of a voice frequency through modulation.*

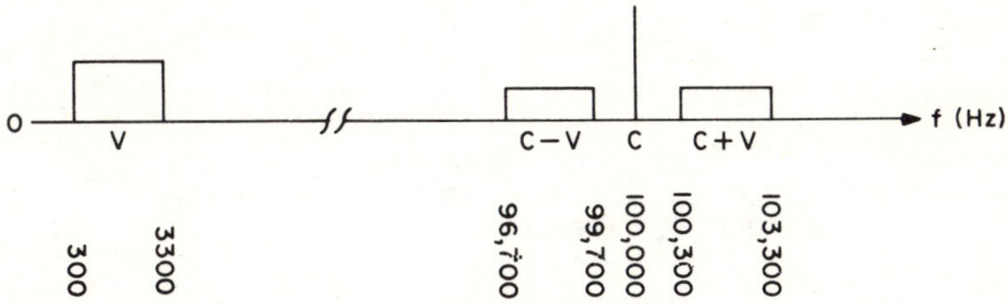


Figure 6-12. *Creation of sidebands with frequency ranges.*

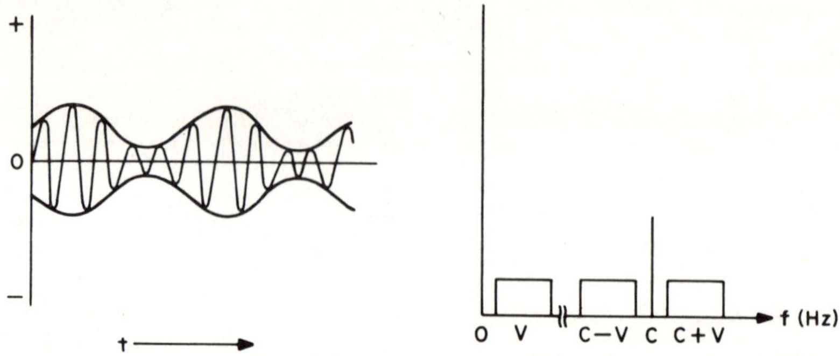


(A) DOUBLE-SIDEBAND AMPLITUDE MODULATION (DSB-AM)

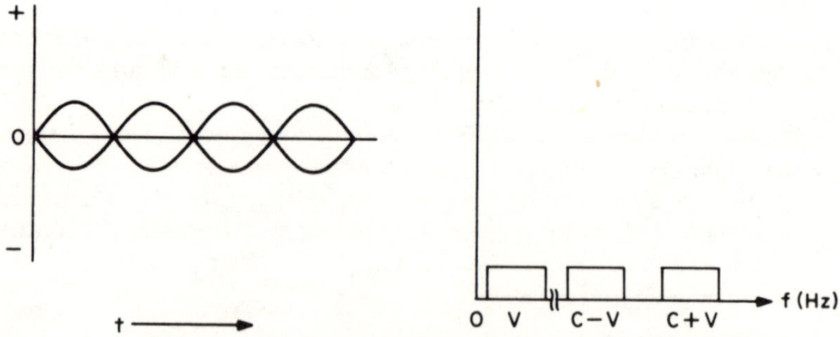


(B) DOUBLE-SIDEBAND AMPLITUDE MODULATION WITH SUPPRESSED CARRIER

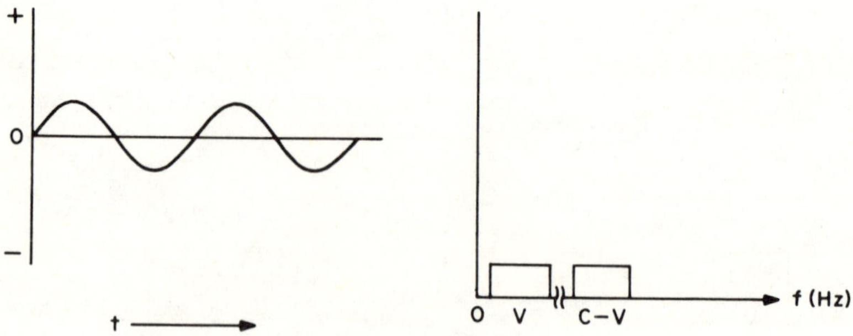


(C) SINGLE-SIDEBAND AMPLITUDE MODULATION WITH SUPPRESSED CARRIER

Figure 6-13. *Several variations of amplitude modulation.*

within certain limits. The telephone voice band, with its approximate limits of 300 and 3300 Hz, is an example. In this case, there is a signal in a bandwidth of about 3000 Hz modulating the carrier frequency, and C + V would be a range of frequencies just as it is in the original voice band, rather than the single 101,000 Hz wave it was above (see Fig. 6-12).

Figure 6-12 was an example of *double sideband* modulation (DSB). Each of the sidebands carried the same information as the original modulating wave, V. But V took up only 3000 Hz of bandwidth, while C and its two sidebands required 6600 Hz (103,300 − 96,700). Since each sideband carries all the information that V has, using only one sideband will thus economize the available bandwidth. Figure 6-13 shows how this is done.

Single-sideband transmission has the advantages of requiring a lower amount of power than does double sideband and of requiring much less bandwidth. If its carrier is suppressed, however, the exact frequency of the original carrier must be reinserted at the receiver or frequency translation may occur; that is, the frequency of the demodulated signal may not be the same as that of the original modulating signal.

Vestigial sideband, used in some data modems and for transmitting television, also uses less bandwidth than DSB. The preservation of the upper portion (vestige) of the lower sideband prevents clipping of the upper sideband and loss of the control frequencies. It also reinforces the important lower signal frequencies (which carry the synchronizing pulses, most of the signal energy, and the major values of the image structure). The main reason a portion of one sideband is partially suppressed is that from a practical standpoint it is nearly impossible for a filter to cut off sharply enough to completely suppress one sideband while leaving the carrier and other sideband intact. (See Fig. 6-14.)

## Angle Modulation

Angle modulation varies the *frequency* of the carrier, unlike all the previous examples, in which the carrier's *amplitude* was varied. The two

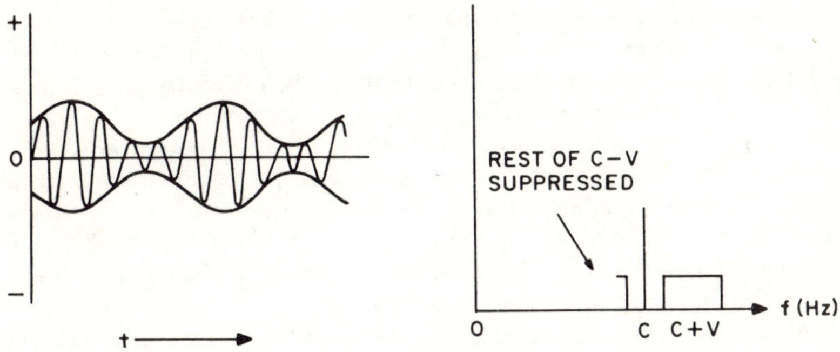


Figure 6-14. *Vestigial sideband (VSB) amplitude modulation.*

types of angle modulation are frequency modulation (FM) and phase modulation (PM).

The Bell System Data-Phone data set 200 series uses FM. Binary d-c polar signals from a business machine are introduced to the data set, where the signals are used to control the frequency output of a multivibrator. In this case, a positive d-c current from 3 to 50 V represents a spacing condition, and negative current of the same value a marking condition. These signals cause the multivibrator to change between two output frequencies, which are amplified, filtered to screen unwanted harmonic frequencies, and sent out on the transmission circuit.

An FM data signal consists of a carrier, C, and sideband frequencies of C ± M, C ± 2M, C ± 3M, and so on, where M is the modulating frequency (see Fig. 6-15). During modulation, energy from the carrier is distributed to the sidebands, reducing the carrier energy. The required bandwidth of an

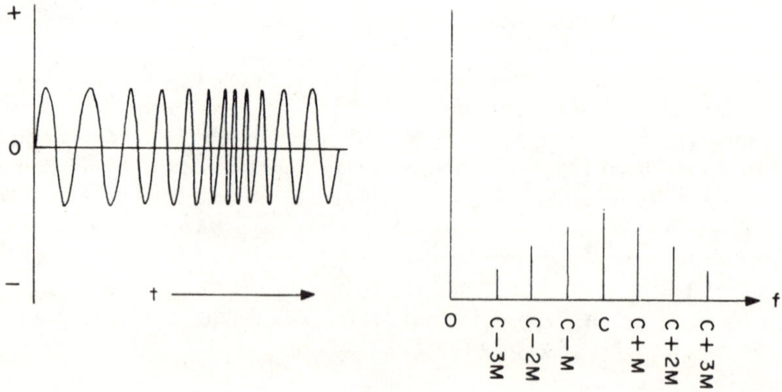


Figure 6-15. *Characteristics of an FM signal.*

FM system is usually greater than that of an AM system and is determined by the amplitude of the modulating wave. An FM signal is said to be fully modulated when it has reached ±75 kHz from the carrier frequency. Since the amplitude of an FM signal does not vary, the transmitter may operate at continuous maximum efficiency. FM receivers produce squarewaves which reproduce the zero crossings, but not the amplitude, of the original signal. The Bell Systems Data-Phone data set 202 receiver produces baseband signals of ± 8 V from the received FM signal. These baseband signals resemble those of the transmitting business machine, and are interpreted as such by the receiving business machine.

Another method of modulating a carrier wave is to arrange for a change in a modulating signal to produce a corresponding change in the phase of the carrier (phase modulation). A simple example of this would be a binary

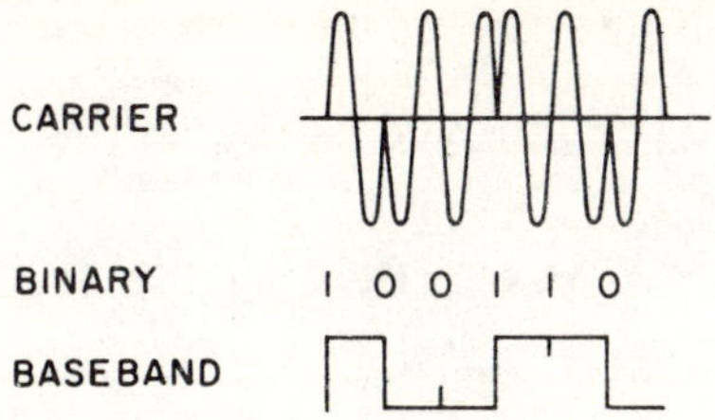


Figure 6-16. *Phase modulation (PM) of a binary signal.*

baseband data signal which, at every change in its state, causes a corresponding 180° shift in the phase of the carrier it was modulating (see Fig. 6-16).

| PAIR | PHASE |
|---|---|
| 00 | 0 |
| 01 | 90 |
| 10 | 180 |
| 11 | 270 |

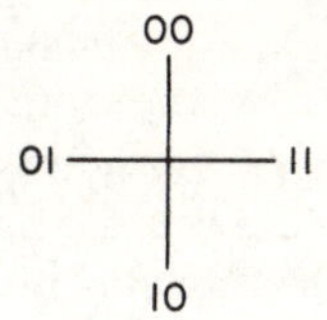


Figure 6-17. *Pairs of binary digits in four-phase modulation.*

Binary PM is called two-phase modulation. Four-phase modulation is merely a phase shift of 0, 90, 180, or 270 degrees, depending on the sequence of values of the modulating quaternary signal. Pairs of binary digits can combine in only four ways: 00, 01, 10, and 11. We can assign each of these pairs a phase value, as seen in Fig. 6-17. Then, we can transmit a carrier that shifts phase for each pair of bits in a sequence of message bits.

A message with bit sequence 0111000110 was divided into the dibit sequence 01 11 00 01 10, which modulated a carrier once per cycle and produced the PM wave shown in Fig. 6-18. This type of modulation, done by

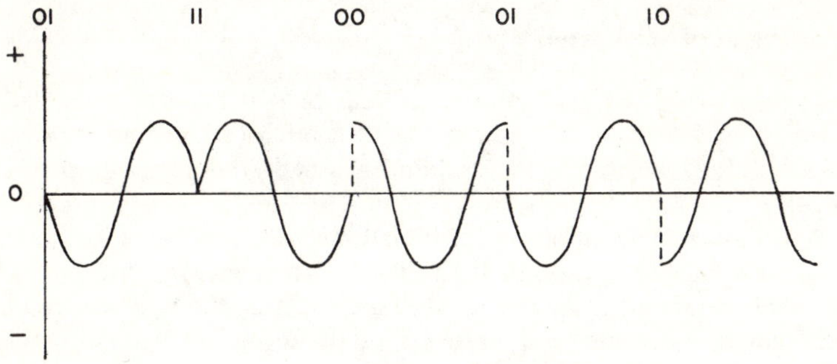


Figure 6-18. *A phase-modulated carrier.*

the Bell System Data-Phone data set 201 series, allows transmission of two bits per cycle of carrier. Actual throughput is considerably less, however, to allow for ample detection time in the receiver.

### Pulse Modulation (PAM, PPM, and PDM)

Although AM and FM are the most popular modulation methods used for data transmission, some systems such as quantized analog data may use some form of pulse modulation. A brief discussion of PM is included here to acquaint the reader with the terms and concepts.

In both AM and FM, the modulating signal has some direct modifying effect on a carrier frequency. In pulse modulation, however, the signal is sampled and encoded in one of several ways. If the sampling were done with less frequency than the rate of change of the signal, significant characteristics of the signal might be missed. To preserve the signal's information content, therefore, the sampling must be done at least twice per cycle of the highest frequency transmitted.

The sampling process determines the instantaneous value of the signal at the time it is sampled. This procedure results in a series of pulses that vary in proportion to the characteristics (for example, amplitude) of the signal. The resulting pulses, depending on the coding scheme used, vary in increments of amplitude, position, or time.

Figure 6-19 compares the three systems of pulse amplitude (quantizing), pulse position, and pulse duration (or pulse width) modulation. The signals resulting from pulse amplitude modulation (PAM) are of constant duration, but vary in amplitude in proportion to the instantaneous value of the source or modulating wave. Pulse position modulation (PPM) signals have constant amplitude and duration, but the time between occurrence of a pulse and the point at which it would occur without modulation is proportional to the value of the source signal at the instant it was sampled. Pulse duration modulation (PDM) produces a signal that varies in duration (width) in proportion to the modulating signal, while remaining constant in amplitude.

### Bit Capacities of Modulated Waves

Theoretically, Nyquist's maximum rate of two bits per cycle may be used as a standard for comparing the efficiencies of various modulating schemes. A bandwidth of N cycles, in other words, theoretically could carry 2N bps.

Double-sideband amplitude modulation would require twice the bandwidth of the baseband modulating signal to carry the same number of bits, since *each* sideband is of the same bandwidth as the original baseband signal. Single sideband and vestigial sideband signals approximate the

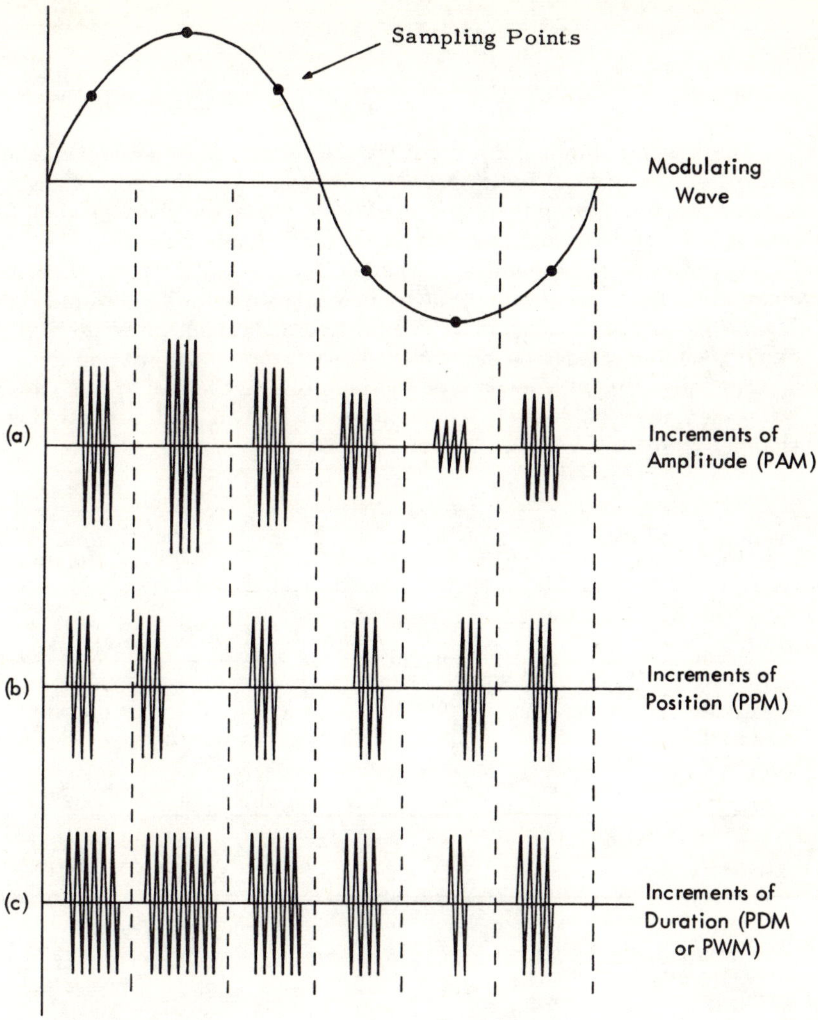


Figure 6-19. *Comparison of (A) pulse amplitude, (B) pulse position, and (C) pulse duration modulation. Sampling rate is six per cycle.*

bandwidth of the modulating signal itself, and so would carry the same number of bits. FM and double-sideband AM usually require about the same bandwidth, so their bit capacity would also compare. Some phase modulation schemes, however, permit the transmission of dibits, and thus actually approach Nyquist's maximum of two bits per cycle of bandwidth.

## MULTIPLEXING

### Introduction

The effect of modulation is the translation of a low-frequency bandwidth from its original location in the spectrum to a location of higher frequency. Multiplexing merely is *repetitive modulation:* the "stacking" of several low-frequency signals into a higher place in the spectrum.

Multiplexing a communications circuit causes it to carry more than one channel at a time. The advantage of multiplexing is that the wide bandwidths of most circuits (such as a pair of wires or a coaxial cable) can be utilized to much greater advantage by being made to carry more than just one signal.

An early attempt to overcome the ratio of one channel to one circuit was called the "phantom." It was invented in 1902 and derived three channels from only two circuits. (See Fig. 6-20.)

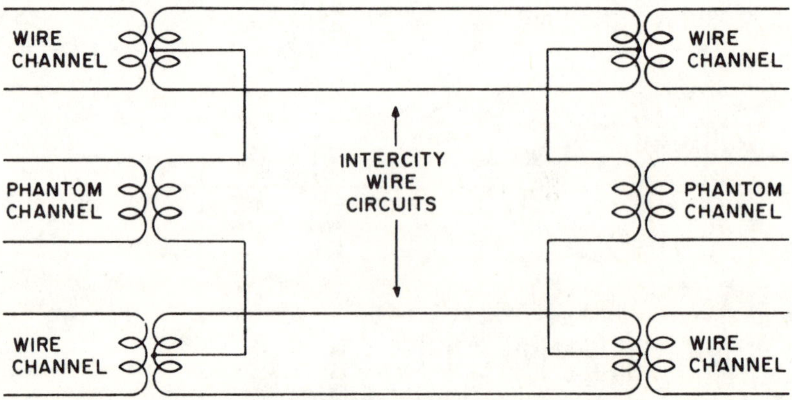


Figure 6-20. *Derivation of third (phantom) channel from two wire circuits.*

The two intercity wire circuits were terminated in electrical networks at each end. In each city, local wire channels (pairs) were electrically coupled to the intercity circuits by use of the network, forming two complete and independent channels.

Each pair of intercity wires was caused to appear electrically as one wire by centertapping the network of each, and coupling electrically a third local channel (the phantom) to the centertap. The receiving end of the circuit retrieved the signal by coupling to the receiving network centertap. Thus two pairs of wires effectively carried the load of three, but had to be kept finely balanced electrically to prevent crosstalk.

**Frequency-division Multiplexing**

Several discrete but simultaneous signals can be imposed on one circuit if each signal occupies a different "slot" or portion of the available circuit bandwidth. Figure 6-21 illustrates lines (signals) 1 through N being multiplexed by frequency division on to one line containing N channels, each of a discrete, narrow frequency range. At the sink, the signals are derived from each channel by filtering, restored to their previous characteristics, and introduced to the outgoing lines.

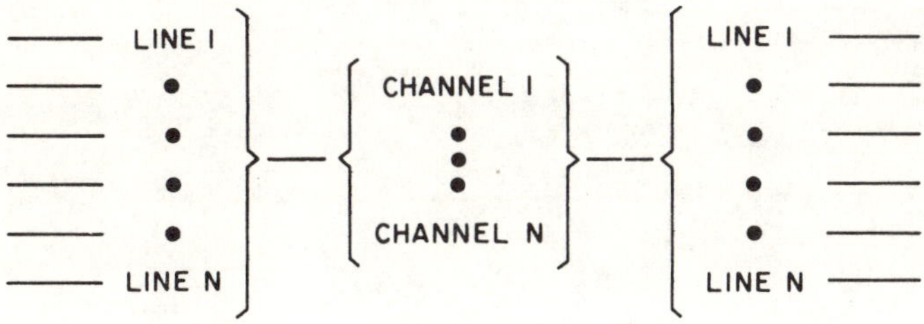


Figure 6-21. *Frequency-division multiplexing.*

Figure 6-22 graphically portrays 12 voice-grade channels ($V_1$ through $V_{12}$), each occupying a frequency range of 300 to 3300 Hz, being multiplexed to discrete slots at a much higher frequency level. Notice that the carriers are placed 8 Hz apart, even though the original voice bandwidths are only 3000 Hz wide. This is to allow a safety buffer between channels to help prevent cross-modulation.

If all 12 voice signals had been imposed on the circuit at their original frequency level, they would have impinged on one another and garble would have resulted. Most circuits have much wider bandwidths than that required by one voice signal, so much more efficient use of the circuit is obtained by spreading the signals over the available bandwidth as described above, and sending them simultaneously. At the destination the signals are reduced to their original voice-frequency range and distributed one to a circuit (line).

Multiplexing a double-sideband carrier is wasteful of available frequency in two ways: there are many buffer zones, and two sidebands are required for each original voice band. To alleviate this, we can send each voiceband as a single sideband with suppressed carrier. The 12 voiceband channels would then appear as in Fig. 6-23. The 1000 Hz buffer zone between each 3000 Hz voiceband channel would be preserved, and there would be no waste of the frequency spectrum. Twelve voiceband (4 kHz) channels arranged in this way are called a *group*.

Earlier, multiplexing was defined as being repetitive modulation. A

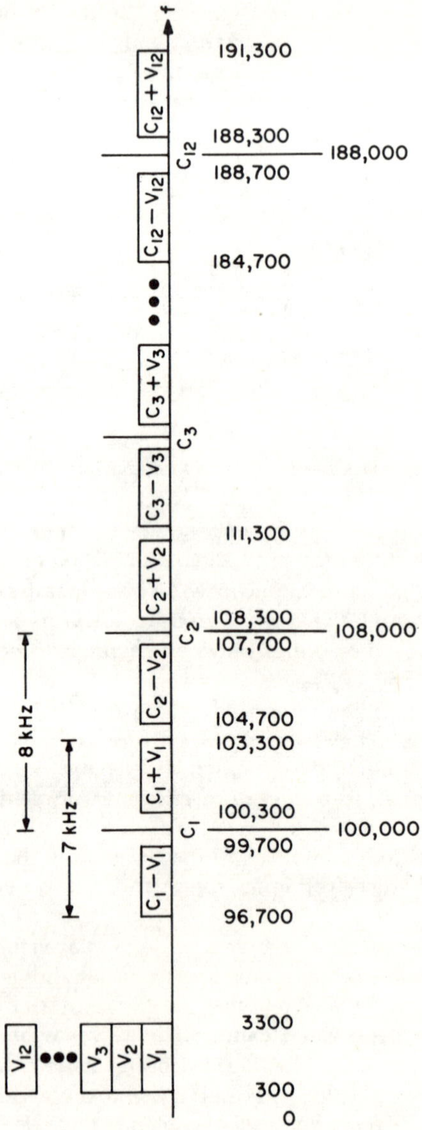


Figure 6-22. *Distribution of voiceband channels within the frequency spectrum with double sideband multiplexing.*

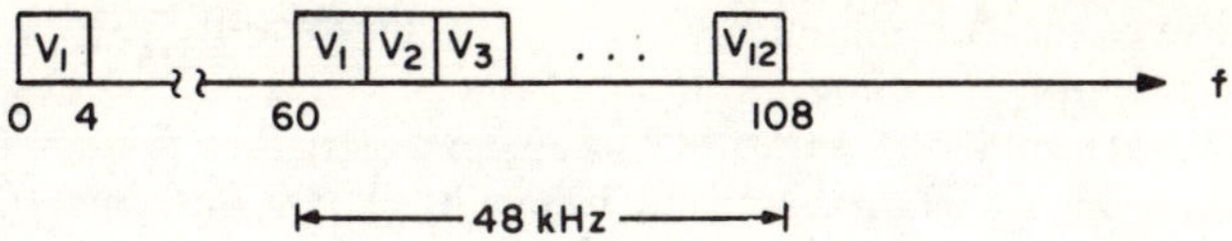


Figure 6-23. *Creation of 12 voiceband channels (one group).*

good example of this is the way "long-haul" telephone systems are multiplexed. Through repetitive modulation, the following systems are obtained:

| | | |
|---|---|---|
| 12 voice channels | = | 1 base group |
| 5 groups | = | 1 supergroup (60 channels) |
| 10 supergroups | = | 1 L1 system (600 channels) |
| 3 L1 systems | = | L3 system (1800 + channels) |

Figure 6-24 portrays this relationship graphically. Table 6-1 provides a summary of most telephone carrier systems in use today.

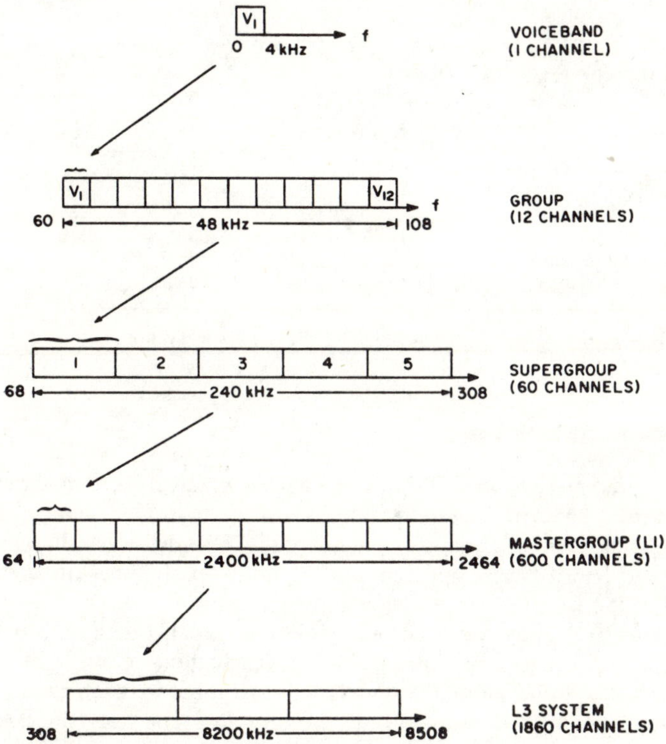


Figure 6-24. *Creation of carrier systems by frequency-division multiplexing (hypothetical).*

Multiplexing techniques are not limited to voice-frequency channels, but may be applied to any frequency ranges. These techniques are often applied to teletypewriter and other low-speed signals so that many of them can be combined and economically transmitted over one voice-grade line. Many multiplexing devices are available to the consumer for this purpose.

TABLE 6-1. SUMMARY OF TELEPHONE CARRIER SYSTEMS

| Carrier System | Trans-mission Medium | Line Frequency Band | Equivalent Telephone Circuits | Long or Short Haul* | Max. mileage |
|---|---|---|---|---|---|
| Single Voice Band | Wire or Cable 1 Pair | 300 - 3300 Hz | 1 | Short | |
| C | Open Wire (1 Pair) | 4.6 - 30.7 Hz | 1 | Short | 500 |
| J | Open Wire (1 Pair) | 36 - 143 kHz | 12 | Short | 800 |
| N | Cable (2 Pairs) | 44 - 260 kHz | 12 | Short | 200 |
| O | Open Wire (1 Pair) | 2 - 156 kHz | 16 | Short | 150 |
| P | Open Wire | 8 - 100 kHz | 4 | Short | 25 |
| ON | Cable | 36 - 268 kHz | 24 | Short | 200 |
| ON/K | Cable | 68 - 136 kHz | 14 | Short | 200 |
| TJ | Radio | 10.7 - 11.7 GHz | 1200 | Short | 200 |
| K | Cable (2 Pairs) | 12 - 60 kHz | 12 | Long | |
| LI | Coaxial Cable | 68 - 2788 kHz | 600 | Long | |
| L3 | " | 312 - 8284 kHz | 1860 | Long | |
| L4 | " | 564 - 17,548 kHz | 3600 | Long | |
| TD-2 | Radio | 3700 - 4200 MHz | 600 | Long | |
| TH | Radio | 5925 - 6425 MHz | 1860 | Long | |

*Short-haul systems vary from 25 to 800 miles, but usually are about 100 to 200 miles.

## Time-division Multiplexing

Frequency-division multiplexing assigns each of several signals to a discrete slot (narrow frequency range) within a particular bandwidth, and transmits them simultaneously on the circuit. Time-division multiplexing, on the other hand, takes representative samples of the several signals, and transmits the samples *sequentially* on the circuit.

Figure 6-25 illustrates five teletypewriter signals being multiplexed by time-division on to one output channel. Each signal consists of six code elements (five information and one synchronization) for each character. The duration of each code element is divided into five time slots (since there are five signals to be sampled). During the first time slot, input channel A is sampled, a signal (voltage, frequency, current, etc.) is detected, and a corresponding signal element is imposed on the output channel. During the

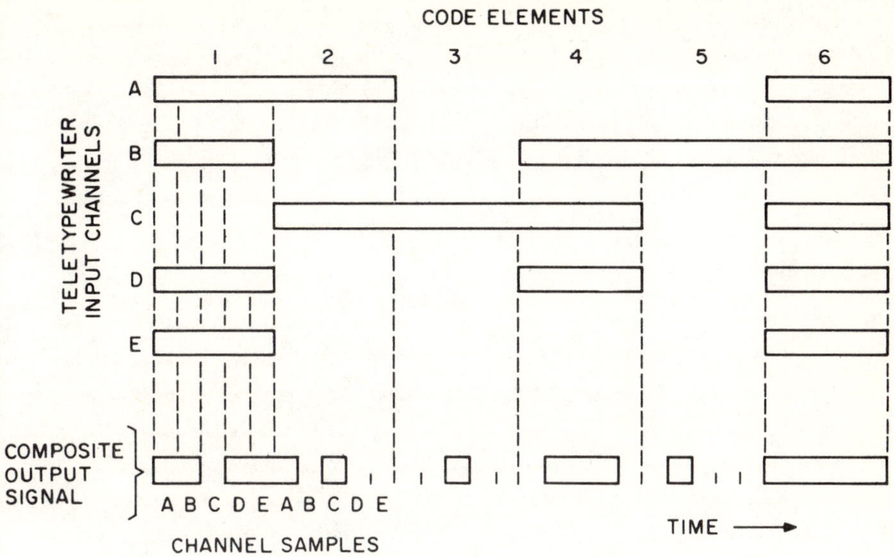


Figure 6-25. *Time-division multiplexing.*

second time slot, input channel B is sampled, and the presence of a signal causes a signal element to be imposed on the output channel. During the third time slot, channel C does not contain a signal, and therefore, no signal element is imposed on the output channel.

At the end of the duration of the first code element, all five input signals have been sampled, and the samples (or other representations) imposed on the output channel. Sampling must be done at a rate high enough to sample each input channel during each code-element duration.

At the sink, the composite signal is demultiplexed by reversing the above sampling technique, and re-creations of the original signals are transmitted to their destinations on individual circuits. Figure 6-26 graphically compares frequency-division and time-division multiplexing.

Another way to think of time-division multiplexing is as a parallel-to-serial conversion process, in which samples of several parallel (or simultaneous) signals are transmitted sequentially on a single higher-speed circuit. A small increment of time is allotted for transmission of a sample from each input signal or channel. The samples are transmitted in sequence until all input signals have been sampled, and then the cycle is repeated. The sampling cycle must be sufficiently fast to insure that the samples accurately reflect the value of the input signal. An analog signal, for example, must be sampled at least twice as fast as its highest frequency if the resultant sample is to be accurate.

Since time-division multiplexing has the ability to reduce the number of channels required to transmit information, it is often economically attractive. Many teletypewriter voice channels can be time-division multiplexed

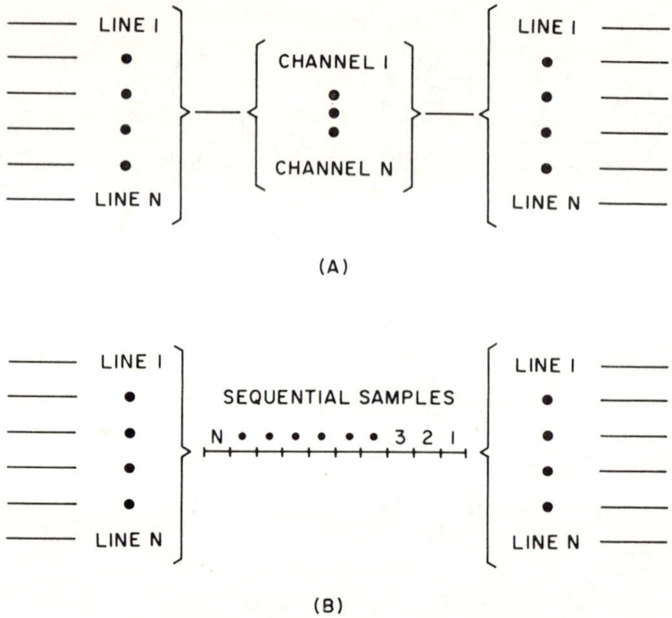


Figure 6-26. *Comparison of (A) frequency-division multiplexing and (B) time-division multiplexing.*

into a greatly reduced number of cross-country or transoceanic channels. Large numbers of meters, alarms, or monitoring devices can be sensed or operated by remote control. Computers and other message-switching devices can use this method to handle large numbers of transmission lines, each operating at its own speed and carrying its own code.

A synonym often used for multiplexing is concentrating. In addition to the economies resulting from reducing the number of required lines, the use of a concentrator can often reduce equipment costs. This is true since equipment costs tend to rise rapidly in proportion to the number of lines terminated in it. It is usually much less expensive to handle a complex, high-speed signal over one line than to terminate simple low-speed signals arriving on several lines.

Concentrator functions can be expanded to include code translation and speed translation, thus resulting in even further cost reductions in the main switching or computing equipment. The net effect of a concentrator with these functions is to make the operation of the remote terminals "transparent" to the computer. In other words, the computer need not be programmed or equipped to handle each code, each speed, or a multitude of other attributes peculiar to individual types of terminals, but may merely process or switch the information and let the concentrator handle the "administrative" details.

Multiplexing techniques are not limited to voice-frequency channels, but may be applied to any frequency ranges. These techniques are often applied to teletypewriter and other low-speed signals so that many of them can be combined and economically transmitted over one voice-grade line. Many multiplexing devices are available to the consumer for this purpose.

## Problems

1. Define: carrier, baseband signal, modulation.
2. What is sampling?
3. Give two purposes for modulation.
4. What are three general classifications of modulation?
5. How do amplitude and angle modulation differ?
6. What are three types of pulse modulation?
7. Define multiplexing, and give its advantages. What, if any, are its disadvantages?
8. Name and describe two kinds of multiplexing.

*CHAPTER 7*

# *Commercial Communications Channels and Services*

## INTRODUCTION

Until the invention of carrier systems, every telephone connection required an individual pair of wires. Today, one pair of wires may carry up to 24 individual telephone channels, and other systems have capacities numbering in the thousands of telephone channels. This chapter will explain the classes, grades, and types of communications media, and relative pricing of the various channels for commercial use.

## CLASSES OF TRANSMISSION MEDIA

### Introduction

Not too long ago, a person making a long-distance (LD) call could visualize his voice flashing along miles of shining copper wire. Today, his conversation may be carried to its destination by as many as six different media classes: open wire, cable, carrier, coaxial cable, radio, or microwave radio. The following paragraphs briefly define these classes of transmission media, each of which is capable of carrying multiple telephone channels.

### Open Wire

This medium was once the most familiar, particularly to those who have taken rides in the country and have seen miles of copper wire suspended by insulated cross-arms on telephone poles. Although many miles of open wire remain in the country, it is rarely seen in urban areas, where it has been replaced by other media.

**Cable**

Telephone cable is an assembly of insulated pairs of wires. A single cable can contain from several to thousands of pairs, each with color-coded insulation for identification, and each pair capable of carrying one or more voice channels. Older cables were encased in lead sheaths, but modern cables have polyethelene sheaths. Cable has largely replaced open wire, especially in the distribution of pairs from central switching offices to businesses and residences.

**Carrier**

In carrier systems the "transmission of (baseband) signals is effected by impressing the signal voltage on a carrier wave having a relatively high frequency. The signal is thus transmitted to its destination by electrical waves whose frequencies are normally more nearly comparable in value to the carrier frequency than to the signal frequency. The basic reason for this procedure in both carrier and radio systems is to make possible the transmission of a number of different signals over the same transmitting medium without mutual interference by placing each signal in a different portion of the frequency spectrum."[1] Carrier is usually transmitted over metallic media, whereas radio (which includes microwave) is usually sent through the air.

**Coaxial Cable**

The following is an excellent explanation of coaxial cable.

> A modern coaxial cable used for communications consists of a number of coaxial conductors called "tubes" or "pipes." Such a cable has as many as twenty tubes, and also contains various other wires, for control, alarm, and maintenance purposes. The cable and the tubes are electrically shielded and protected in various ways.
>
> Each tube or pipe is made up of copper tubing, usually 3/8 of an inch in diameter, that has a smaller copper wire running through it. The wire is held exactly in the center of the copper tubing by small disc-like insulators spaced about one inch apart. Since the tubing and wire both have the same center axis, this arrangement is called "coaxial."[2]

**Radio**

Each of the above means of transmission has used a metallic medium. Various circumstances, such as mountainous terrain, urban development, or very large volumes of traffic, lead to the consideration of radio as a transmission medium.

[1] *Principles of Electricity,* AT&T Long Lines Dept., N.Y., 1961.
[2] *Communication News,* October, 1965, p. 24.

It will suffice for our present purposes to say that radio systems perform practically in the same way as the carrier systems mentioned above; many baseband signals are "stacked" in a higher portion of the frequency spectrum by multiplexing, and are transmitted to a distant point along a single path. The path in this case, however, is a radio signal in space rather than the metallic circuit of a carrier system.

### Microwave Radio

Microwave is a special case of radio transmission, in that it occupies the frequency spectrum between 890 megahertz and 30 gigahertz, although frequencies above 16 gigahertz are unassigned for specific uses. At these frequencies, radio waves require special equipment during transmission and reception, and they are limited to line-of-sight transmission. The attractive features of microwave are the capacity to transmit large quantities of information, and to do so without the necessity of stringing miles of connecting circuits between points.

### Laser

Intense experimentation is being conducted toward the practical application of the laser (Light Amplification by Stimulated Emission of Radiation) to communications problems. It has the advantages of tremendous information capacity and of not occupying the already crowded r-f spectrum. Like microwave, laser beams travel line-of-sight routes, but their use of optical frequencies create unique problems that must be given practical solutions.

## TYPES OF COMMUNICATIONS CIRCUITS

### Introduction

Any of the transmission media discussed above can be used in various ways to move information. The choice depends upon such factors as distance, volume, economy, bandwidth, and so forth. In most cases, however, the type and grade of circuit to be used are of more concern than the class of media through which the circuit will pass.

The *types* of circuits are simplex, half-duplex, and full-duplex, and will be discussed below. A description of *grades* of channels (narrowband, voiceband and wideband) will then follow.

### Simplex Circuit

The term simplex used in reference to a communications circuit has two possible meanings. The first refers to a circuit that has a transmitter at one

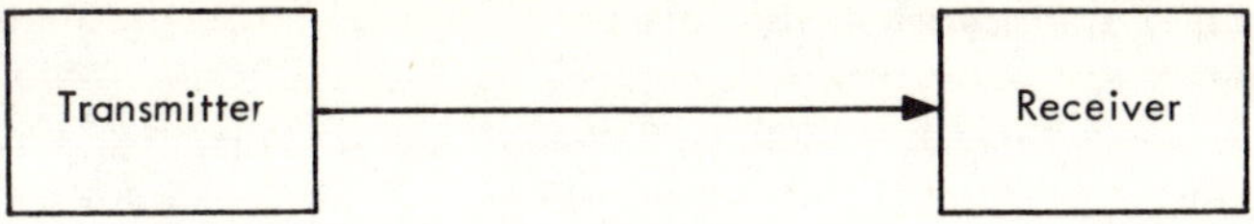


Figure 7-1. *A simplex circuit.*

end of a channel and a receiver at the other and is thus restricted to one-way transmission. The term can even apply to the channel itself, since the presence of echo suppressors, repeaters, and other equipment can permit traffic in one direction while inhibiting it in the other. (See Fig. 7-1.)

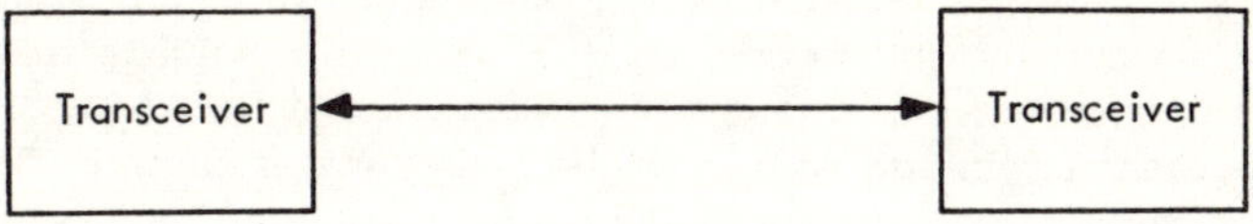


Figure 7-2. *A half-duplex circuit.*

The second meaning of simplex is a channel, circuit, or system which carries only one message at a time. In other words, it has not been multiplexed.

**Half-duplex Circuit**

A half-duplex circuit allows transmission in both directions, but in only one direction at a time. An example is the common TWX connection, which allows an exchange of information between two teletypewriters, but only one may be transmitting at any given time. (See Fig. 7-2.)

**Full-Duplex Circuit**

A full-duplex circuit permits transmission in *both directions simultaneously.* Such a circuit (often referred to simply as duplex) must be multiplexed to prevent the opposing signals from interfering. This is sometimes done by assigning frequencies in the upper half of the channel's

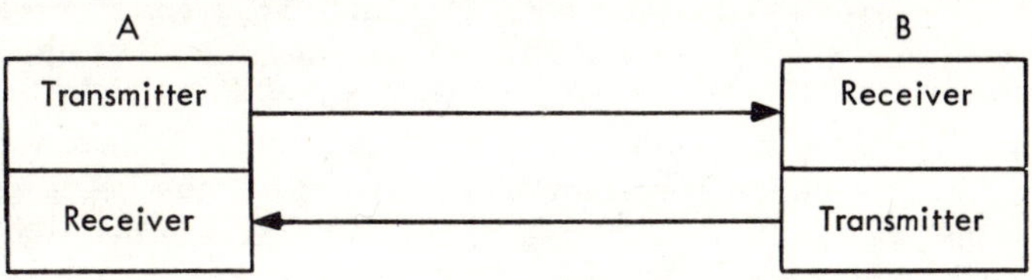


Figure 7-3. *A full-duplex circuit.*

bandwidth to transmitter A and receiver B, and the lower frequencies to transmitter B and receiver A (see Figure 7-3). A more frequent method of achieving full-duplex capabilities is to use two pairs of wires (two channels), one for each direction of transmission (that is, two simplex circuits). This allows faster speeds than the subdivision of a single channel, but costs more due to the additional facilities or circuit conditioning required.

## COMMERCIAL COMMUNICATION FACILITIES

### Common Carriers and Tariffs

Common carrier, when used in conjunction with communications, refers to a company which dedicates its facilities to a public offering of a range of communications services and which is subject to regulation. The communications common carriers in the United States comprise Western Union Telegraph Company and many independent telephone companies. The largest of these telephone companies is American Telephone and Telegraph which, with its subsidiary Bell companies, provides about 85 percent of our telephones. The second largest telephone company, General Telephone (a Sylvania subsidiary), provides about 8 percent. The remaining 7 percent are provided by about 2700 independent telephone companies.

All communications common carriers are regulated by published tariffs which specify rates for units of equipment, facilities, or types of service provided by the carrier. Intrastate tariffs are usually filed by the carrier with a state regulatory body, such as a public utility commission or a corporation commission. Intrastate tariffs apply only to those services that are provided wholly within the boundaries of the regulated state. Interstate tariffs are filed with the FCC by Western Union and AT&T for services that cross state boundaries.

### A Note of Caution

The descriptions of commercial communications facilities and services discussed in the balance of this chapter should be considered as representative only, since they are administered under tariffs and specifications which are subject to change. Interstate rates and specifications have been used whenever possible. Intrastate rates vary from state to state, and usually are somewhat higher than interstate rates. The appropriate common carriers should be contacted for verifications of rates, specifications, and usage regulations.

Because detailed descriptions and costings of every common carrier offering would be a duplication of published tariffs and other sources, most of the following facilities will be described in general terms. Clarifying details will be added where appropriate.

**Exchange vs Private (Leased) Facilities**

Exchange facilities comprise channels and equipment used by the communications common carriers to provide service to the general public. A good example is the switched voice network, in which the switching equipment, connecting channels, and terminal equipment are used by a great number of users in succession and usually without interference.

In contrast, private (leased) facilities are dedicated to the use of one subscriber exclusively. Rather than paying on a per-call basis as with exchange facilities, the customer who leases facilities pays a flat monthly fee and is entitled to use them to their full capacity (within certain restrictions as to the type of signals which may be transmitted). Private facilities range from two-point telephone lines to nationwide switched voice and data systems such as Common Control Switching Arrangement (CCSA).

## NARROWBAND FACILITIES

The commercial facilities discussed below are grouped by grade of channel: narrowband, voiceband, and wideband. In each case, a brief description of each channel grade will be followed by a general description of some of the services provided by common carriers for use in conjunction with that grade. Common carriers do not usually offer bandwidths as such, but instead they define the characteristics or capabilities of an offering. They may, however, refer to "equivalent voice channels," which is a convenient standard, or state a nominal maximum bit or baud rate which may be expected of a facility.

**Channels**

Narrowband channels are those which have less bandwidth, and thus less information capacity, than standard voice channels. For some time, the only common narrowband channel was of teletypewriter grade, but recently some channels of somewhat larger bandwidths, referred to as subvoice grade, have been offered.

Commonly, 12 standard teletypewriter-grade channels can be derived from one voice channel, and each of the 12 can carry up to 75 bauds. Two of the newer subvoice channels can carry 150 and 180 bauds, respectively, and each occupies one-sixth of a voice channel. Another occupies one-half of a standard voice channel.

**Telemetering**

Broadly, telemetering is the transmission of measurement results. The telemetering signals from the guidance system of a missile in flight may require relatively wide bandwidths. Other measurement information, however, such as the setting of a dial, the flow of fluid through a pipe, or the temperature of a chemical process, requires only a small portion of the frequency spectrum to be transmitted at satisfactory speeds. There is one channel offering, in fact, that has only half the capacity of a telegraph channel.

### Teletypewriter

Narrowband channels provide the vehicle for both private-line and exchange teletypewriter service. The private-line services may be two-point or multipoint and either manually or automatically switched. Examples of switched private-line systems include the Bell System's 83B2 and Western Union's comparable Plan 115A. Private-line teletypewriter systems are referred to by the use of the acronym TWPL. Exchange teletypewriter systems include Bell TWX (Teletypewriter Exchange service) and Western Union TELEX.

### Low-Speed Data

Data can be transmitted over narrowband facilities at speeds up to 180 baud. Teletypewriter channels accommodate signals up to about 75 baud. Both exchange and private-line narrowband facilities can be arranged for data transmission.

## VOICEBAND FACILITIES

### Channels

The voiceband channel, because of its prevalence, economy, and convenient bandwidth, has become the standard facility for data communications, and is the most used. A telephone installation anywhere is capable of accepting data transmission if the proper terminal equipment is provided. And, dollar for dollar, it would be difficult to find a more economical communications medium than the ubiquitous telephone line. The nominal 4 kHz bandwidth of a voice channel allows relatively large amounts of data to be transmitted within a reasonable amount of time.

At the end of this chapter is a table of channel characteristics which lists many varieties of channels available in the voiceband. These channels range in quality from an ordinary dialed connection through the exchange network to specially conditioned channels that will accept data signals at speeds up to 2400 baud.

### The Dialed Network

The dialed telephone network provides ordinary telephone service over a nationwide system of switched lines. It is often commonly referred to as the Direct Distance Dialing (DDD) network, the public switched network, the public telephone system, and so forth. Because of the ease of installing dialed telephone service nearly anywhere, and of connecting data communication devices to this service, it has become a common method of low and medium-speed data transmission. Charges for using the network for data are the same as for voice and are based on a combination of distance and time.

The dialed network may be used to convey many forms of information, such as digital data, teletypewriter signals, handwriting, facsimile, visual displays, tone transmissions, and the like, which are discussed below.

### WATS (Wide Area Telephone Service)

For those who use the telephone network a great deal, a billing arrangement called WATS has been established. Full-time WATS service provides for outgoing station-to-station dialed calls to specified geographical areas. Any number of calls can be made from one telephone for a flat monthly rate. A second class, measured WATS, provides the same service for up to 15 hours of use per month at a reduced monthly cost. Time over 15 hours is charged by the hour.

WATS calls can be made to any of seven designated bands, each one including progressively more telephones. These bands, or calling areas, are made up along state boundaries in most cases, and were developed by combining the total number of telephones and of square miles of each state into a measure of service availability.

The first band, for any given state, is all contiguous states or 10 percent of the total interstate service available, whichever is greater. The second band includes those nearest states which comprise about 20 percent of the total interstate service available. The succeeding bands include about 40 percent, 60 percent, 80 percent, and 100 percent, respectively, of the total available interstate service. Several states have very large areas or numbers of telephones, and have been divided into separate sections: New York, Pennsylvania, Ohio, Michigan, Illinois, Texas, and California.

A company subscribing to full-time Band 4 WATS would be entitled to as many calls of any length that could be placed over the telephone connected to the service. The calls could be placed 24 hours per day to any telephone in the area covered by Band 4, and to any phone in Bands 3, 2, and 1. A subscription to Band 1 allows access only to those phones in Band 1. A Band 6 subscription permits calls to phones in all bands except Band 0, which is the home state (or partial state). Band 0 calls can be made to any point within the home state (or area), but no interstate calls are included.

Thus, if total coverage, including the home state, were desired, both Band 6 and Band 0 service would have to be arranged.

**Common Control Switching Arrangement (CCSA)**

The need for a more economical way to handle many, many calls to many, many places has presented itself within the last few years. To take advantage of the economies of Telpak and, at the same time, permit the temporary interconnection of many scattered points, the Bell System offers a Common Control Switching Arrangement. This, in effect, is a private, long-distance, dialed network. The intercity telephone circuits are priced under private-line or Telpak tariffs, whichever is the more economical. The switching is done by crossbar-type central office equipment (see Chap. 8) allocated to the specific service network.

**Private-Line Service Costs**

Private lines leased from common carriers are charged on the basis of their length in airline miles. To assure uniformity in the measurement between private-line terminals, a grid of vertical and horizontal (VH) coordinates have been established across the United States. Every city is

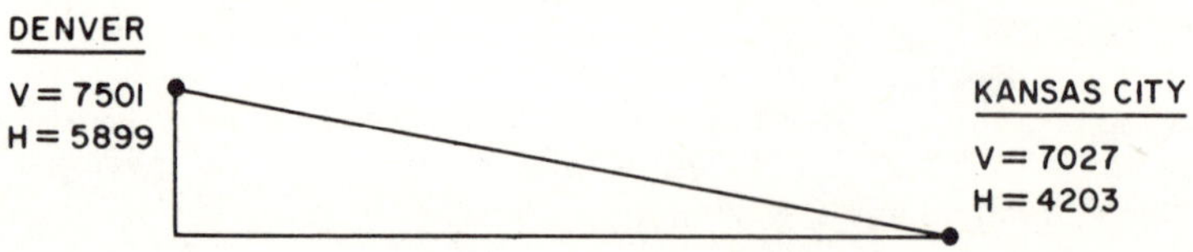


Figure 7-4. *Calculation of private-line costs between two cities.*

identified by a pair of VH coordinates, and the distance between any two cities is found by calculating the hypotenuse of a right triangle of which the cities are located on the two smaller angles (see Fig. 7-4 and Tables 7-1 to 7-3).

The above calculations were for point-to-point (two-point) private lines. Multipoint private-line costs are approached exactly the same way, but introduce the requirement for determining the shortest total distance between all points to be connected. Figure 7-5 indicates several ways that multiple points could be interconnected. Only the fourth solution is the most economical.

The mileage costs of each leg of a multipoint circuit must be calculated as if it were a single system, that is, the mileages of each leg cannot be added together to take advantage of the cheaper long-haul rates.

**TABLE 7-1. CALCULATING PRIVATE-LINE DISTANCE AND COSTS**

| | Location | V | H |
|---|---|---|---|
| (1) List VH coordinates for two cities (see Table 7-2 or FCC Tariff 245) | Denver, Colo. | 7501 | 5899 |
| | Kans. City, Mo. | 7027 | 4203 |
| (2) Subtract | | 474 | 1696 |
| (3) Square the difference | | 474 | 1696 |
| | | 226576 | 2876416 |
| (4) Add both products | | | 226576 |
| | | | 3102992 |
| (5) Divide the result by 10 and take the square root | 5 5 7 miles<br>310299.2 | | |
| (6) Determine PL costs | First 250 mi @ $2.020 = $505.00 | | |
| | Next 250 mi @ 1.717 = 429.25 | | |
| | Next 57 mi @ 1.616 = 92.11 | | |
| | $1026.36 | | |
| (7) Add chan. term. costs | Two @ $12.50 = 25.00 | | |
| (8) Add local channel costs | Two @ $12.50 = 25.00 | | |
| (9) Total monthly PL costs | $1076.36 | | |

See Tables 7-2 and 7-3 for details. Charges shown are for example only.

## DATEL (Overseas Service)

DATEL Service is an offering by three international carriers to provide overseas transmission facilities of a wider bandwidth than telegraph channels on a call-up basis. The service provides data or high-speed telegraph capabilities coordinated with direct overseas voice connections. Facsimile and voice-only calls are not provided by the foreign administrations; DATEL is a voice-coordinated record service.

Interconnection with DATEL Service can be provided through Western Union Broadband Exchange Service, which connects the premises of their customers with the international carrier at the gateway cities. The interna-

**TABLE 7-2. VERTICAL AND HORIZONTAL (VH) COORDINATES FOR VARIOUS U.S. CITIES**

| City | Coordinates | | City | Coordinates | |
|---|---|---|---|---|---|
| | V | H | | V | H |
| Phoenix | 9135 | 6748 | Little Rock | 7721 | 3451 |
| Los Angeles | 9213 | 7878 | Tulsa | 7707 | 4173 |
| San Diego | 9468 | 7629 | Oklahoma City | 7947 | 4373 |
| Oakland | 8486 | 8695 | Amarillo | 8266 | 5076 |
| San Francisco | 8492 | 8719 | Omaha | 6687 | 4595 |
| Denver | 7501 | 5899 | El Paso | 9231 | 5665 |
| Washington, D.C. | 5622 | 1583 | Salt Lake | 7576 | 7065 |
| Kansas City, Kan. | 7028 | 4212 | Portland | 6799 | 8914 |
| Kansas City, Mo. | 7027 | 4203 | Spokane | 6247 | 8180 |
| Dallas | 8436 | 4034 | Seattle | 6336 | 8896 |

**TABLE 7-3. CALCULATING PRIVATE-LINE COSTS**

| Configuration | | Notes | Chg. per Mo. | Inst. Chg. | Billed by |
|---|---|---|---|---|---|
| Subscriber (KC) | Set | Type of signaling and supervision should be specified. Premise charges determined by extent of equipment used. | | | Local Tel. Co. |
| | Local Channel | Tel. set included in local channel charge. | $12.50 | $10.00 | |
| Tel. Co. Central Office (KC) | Channel Terminal | | 12.50 | | AT&T |
| | Inter-State PL | Also called "interexchange channel." If intrastate, billed by local Tel. Co. | 1026.36 | | AT&T |
| Tel. Co. Central Office (Denver) | Channel Terminal | | 12.50 | | AT&T |
| | Local Channel | Also called "local loop." Tel. Set included in local channel charge. | 12.50 | 10.00 | |
| Subscriber (Denver) | Set | Premise charges determined by extent of equipment used. Type of signaling and supervision should be specified. | | | Local Tel. Co. |

Charges shown are for example only.

tional carriers absorb the domestic portion of the usage charges on calls established through Broadband, TWX, Telex, or other wire service.

Each Western Union subscriber desiring DATEL Service must have a Schedule II (4 kHz) Broadband subscription plus the international data set (modem). The usual Schedule II fixed monthly ($30) and installation ($25) charges will be applicable plus the charge for the appropriate international data set.

The data sets are serial in operation (see Chap. 9); parallel international data sets are not available. The synchronous data sets operate at the fixed speeds of 600 or 1200 bps, and the speed may be changed by the patron from call to call. The asynchronous sets operate at the variable rate of 0 to 1200 bps.

It is anticipated that many users with DATEL requirements will also have domestic data requirements. Since the international data set is not compatible with the domestic data set, such users should be encouraged to consider two separate data systems. As an alternative, users may have a single Broadband installation, but two data sets, one domestic and one international, with a manual switch to permit alternate use. In most such cases, the user will utilize a different item of data transmission hardware for domestic as compared to the international traffic.

For those customers who may wish to utilize one piece of data equipment for all their traffic, both domestic and international, two manual switches will be required.

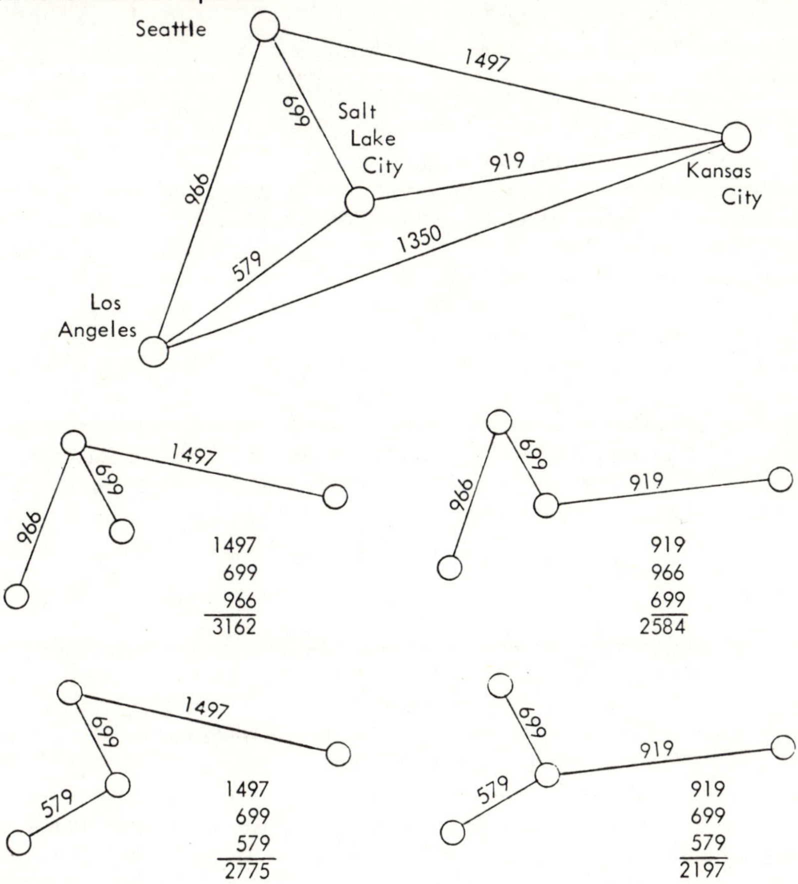


Figure 7-5. *Calculation of multipoint line configuration.*

A data transmission speed of at least 600 bps is assured by the international carriers and their foreign connections. Transmission speeds of 1200 bps will be possible whenever a suitable facility is available at the overseas location.

A wide variety of high-speed data transmission equipment is available from various suppliers of paper-tape, punched-card, magnetic-tape, and computer apparatus for use on DATEL. Most existing domestic data communications apparatus can be adapted for DATEL. Western Union's 1050 wpm equipment has not been approved for DATEL use by foreign administrations.

Subscribers and their overseas correspondents should agree on the type and speed of data apparatus to be used to insure compatibility. Generally speaking, the customer may obtain assistance in this area by contacting one of the international carriers.

## Operation

To place a DATEL call, the domestic subscriber calls the international carrier of his choice at the appropriate gateway city. The international carrier will establish the overseas connection, via voice, after which the domestic subscriber can exchange voice communications to obtain synchronization of both terminals. In receiving a call from overseas, normal data transmission practices are used.

Since the re-ring feature is not operable on DATEL, there is no way to recall the distant subscriber to the voice mode while data is being transmitted. Subscribers with long transmissions may circumvent the lack of this feature by prearranging a procedure with the distant correspondent whereby the data transmission is halted periodically and both parties return to the voice mode to confirm that the data is being received satisfactorily.

DATEL generally operates on a half-duplex basis, but full-duplex may be available to selected countries.

## Billing

The domestic Broadband subscriber will be billed by Western Union for the normal fixed monthly charges for the basic Schedule II Broadband subscription as well as the international data set. The DATEL usage charges will be billed to the domestic subscriber by the international carrier.

DATEL users with AT&T Data-Phone and Dataspeed equipment may call the international carriers (prepaid) and send data at high speeds for relaying overseas at DATEL rates. In these cases, the user pays both the domestic long-distance charge and the DATEL charges and experiences the delay occasioned by the tape relaying and, additionally, is not in direct contact with his overseas correspondent.

### Other Voiceband Services

The list of services available over voiceband facilities is rapidly growing longer. In addition to data, services presently in use include facsimile, handwriting, and visual display terminals. Facsimile (FAX) is the transmission of pictures, maps, drawings, and the like. The original image is scanned by the transmitter, converted to signals suitable for transmission, and reconstructed by a receiver on some form of paper.

Handwriting is translated to a pair of varying tones as the words are written on the paper of the transmitter. These tones can be carried by voiceband facilities to the receiver, where they cause a pen to move over paper and reproduce exactly the images created by the writer at the transmitter. Provision is also made for advancing the paper automatically at both machines simultaneously. Distance is no deterent; handwriting can be sent across the country with the same ease as to a nearby location.

Visual display terminals may take many forms, from a small passive device which slaves on a remote master, to those with television-sized screens and keyboards a local operator can use to communicate with a distant computer. A person using one of many such terminals can converse directly with a computer. Computer utilities are putting visual-display terminals into wide-spread use.

## WIDEBAND FACILITIES

### Channels

Wideband channels are those whose bandwidth exceeds that of the standard 4 kHz voice channel. Commercially available wideband services range from 5 kHz to one million kHz, and are used for radio programs, TV, multiplexed voice channels, voice, and so on.

Examples of wideband channels available commercially are the Telephone Company TELPAK and the Western Union Broadband Exchange Service.

### TELPAK

TELPAK is a leased-line wideband service offered by the Bell Telephone System to permit high-volume communications between given locations. Each TELPAK channel can be used as a single high-speed medium, or can be divided into 60 or 240 independent voice channels for data or voice transmission, according to the subscriber's requirements.

Telephone channels derived from TELPAK services may be used separately for voice or data communications (during the day, perhaps) or may be combined to form one wideband channel capable of high-speed data or facsimile transmission (maybe at night when the voice channels are not used). In addition, each voice channel may be further subdivided into as many as 12 teletypewriter-grade channels if desired. Detailed pricing of TELPAK services can be found in FCC Tariff 260.

## Broadband Exchange Service

Broadband Exchange Service is a Western Union offering designed to provide industry with the means to transmit data and administrative communications at speeds from 180 to 2400 bauds between selected circuits. The service is offered on a metered-time basis with automatic customer-to-customer dialed connection. The terms and conditions under which this service is furnished is detailed in FCC Tariff 246. Cities served by the Broadband Exchange Service are listed in Table 7-4.

The initial Broadband offering was inaugurated in September, 1964, and provides 2 kHz and 4 kHz service only between the following cities; 8 kHz, 16 kHz, and 48 kHz bandwidths will be available in the near future.

TABLE 7-4. CITIES SERVED BY BROADBAND EXCHANGE SERVICE

| | | | |
|---|---|---|---|
| Detroit | New York | Houston | Cincinnati |
| Kansas City | Los Angeles | St. Louis | Boston |
| San Francisco | Cleveland | Atlanta | Buffalo |
| Washington | Dallas | Baltimore | Philadelphia |
| Chicago | Denver | Pittsburgh | Minneapolis/St. Paul |

Subscriber-owned equipment, with the exception of data sets, may be used. Data sets, where required, must be provided by Western Union.

Advantages listed by Western Union for Broadband Exchange Service include: (1) automatic switching of circuits; (2) alternate voice/data; (3) analog or digital data; (4) four-wire circuits; (5) customer-controlled choice of 2 kHz, 4 kHz, (and soon) 8 kHz, 16 kHz, and 48 kHz channels; (6) low noise level; (7) charges on time-usage basis; (8) tone signaling rather than bell ringing; (9) full-duplex operation; (10) initial bit rates from 180 to 2400 per second; (11) high dependability; (12) wide variety of terminating equipment (voice, teleprinter, data, facsimile, etc.).

## Choosing Communication Services

Although the services described above do not include all those available today, they exemplify the various types that are generally used to satisfy most communication requirements.

At first, the proper choice of services to solve a communication problem may appear confusing. Assuming the problem has been properly defined, however, there is a basic set of simple rules of thumb used by Bell System consultants for clarification. Using volume and distribution of calls (voice or

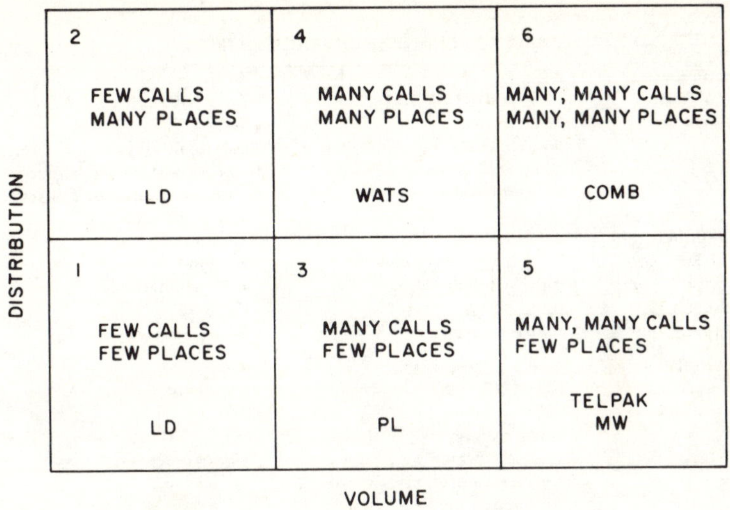


Figure 7-6. *Calling volume and distribution.*

data) to distant points, a customer can locate his situation on a chart (see Fig. 7-6) and get an understanding of the type(s) of services he may require.

Situation one, which has few calls to few places, may be solved by using ordinary long-distance telephone service. Long-distance service also may solve the case of few calls to many places. When many calls are made to just one or a few places, however, a private line or other dedicated service may be more appropriate. WATS may be the solution to situation four, where many calls are placed to many places. If a high volume of calls are made to only a few places, however, a service such as TELPAK or a private microwave system should be considered. The sixth situation calls for much more analysis of the problem and investigation of possible solutions. When many, many calls are placed to many, many places, an integrated combination of services may have to be provided to give a satisfactory system.

## CHANNEL CHARACTERISTICS

Table 7-5 lists information concerning commercial communications channels provided by the common carriers. The characteristics given are the best information available at the time of this printing, but should be confirmed with the appropriate carrier before use. Also, space limitations prohibit the inclusion of amplifying information which is available from published tariffs and material furnished by the carriers. Because of these restrictions, the charts should be used for comparative or initial uses only and not as a basis for final systems design.

**TABLE 7-5. CHANNEL CHARACTERISTICS**

| | | |
|---|---|---|
| Common Carrier | AT&T | AT&T |
| Channel Desc. | Type 1001 (15 Pulse) | Type 1002 (60 & 75 WPM TTY) |
| FCC Tariff No. | 260 | 260 |
| Uses | Remote metering, supervisory control and miscellaneous signaling. | Morse, teletypewriter, data, remote metering, supervisory control, and miscellaneous signaling. |
| Band (+Bandwidth) | Subvoice | Subvoice |
| Type of Service | 2 - point or multipoint | 2 – point or multipoint |
| Operation Mode | Half- or full-duplex | Half- or full-duplex |
| Baud Rate | Up to 30 | Up to 55 |
| Termination | 2-wire or 4-wire | 2-wire or 4-wire |
| Common Carrier | AT&T | AT&T |
| Channel Desc. | Type 1005 (100 WPM TTY) | Type 1006 (Sched. 3A) |
| FCC Tariff No. | 260 | 260 |
| Uses | Morse, teletypewrite, data, remote metering, supervisory control, and miscellaneous signaling. | Mod. 35 TTY and various business machines. |
| Band (+Bandwidth) | Subvoice | Subvoice |
| Type of Service | 2 - point or multipoint | 2 - point or multipoint |
| Operation Mode | Half- or full-duplex | Half- or full-duplex |
| Baud Rate | Up to 75 | Up to 150 |
| Termination | 2-wire or 4-wire | 2-wire or 4-wire |
| Common Carrier | AT&T | AT&T |
| Channel Desc. | Type 2001 | Type 3001 |
| FCC Tariff No. | 260 | 260 |
| Uses | Full-period private-line voice (not used for data). | Same as private-line telephone. (No channel conditioning.) |
| Band (+Bandwidth) | Voice (up to 3 kHz) | Voice (300 - 3000 Hz) |
| Type of Service | 2 - point or multipoint | 2 - point or multipoint |
| Operation Mode | Half- or full-duplex | Half- or full-duplex |
| Baud Rate | Up to 1200 | Up to 1200 |
| Termination | 2-wire or 4-wire | 2-wire or 4-wire |
| Common Carrier | AT&T | AT&T |
| Channel Desc. | Type 3002 (Sched. 4 Type 4 Data) | Type 3003 (Sched. 4 Type 4A Data) |
| FCC Tariff No. | 260 | 260 |
| Uses | Voice and/or Data. Same as Private-line Telephone Service. | Voice, Data, or Alternate Voice and Data. |
| Band (+Bandwidth) | Voice (not specified) | Voice (300 - 2700) |
| Type of Service | 2 - point or multipoint | 2 – point or multipoint |
| Operation Mode | Half- or full-duplex | Half- or full-duplex |
| Baud Rate | Up to 1200 | Up to 1200 |
| Termination | 2-wire or 4-wire | 2-wire or 4-wire |
| Envelope Delay | | Less than 1000 μs between 1000 and 2400 Hz. |
| Attn. with Freq. | | 1000 - 2400 Hz -1 to +3 dB<br>300 - 2700 Hz -2 to +6 dB |

**TABLE 7-5. CHANNEL CHARACTERISTICS (Cont.)**

| | | |
|---|---|---|
| Common Carrier | AT&T | AT&T |
| Channel Desc. | Type 3004 (Sched. 4 Type 4B Data) | Type 3005 (Sched. 4 Type 4C Data) |
| FCC Tariff No. | 260 | 260 |
| Uses | Voice, Data, or Alternate Voice and data. | Voice, Data, or Alternate Voice and data. |
| Band (+Bandwidth) | Voice (300 - 3000) | Voice (300 - 3200) |
| Type of Service | 2 - point or multipoint | 2 - point or multipoint |
| Operation Mode | Half- or full-duplex | Half- or full-duplex |
| Baud Rate | Up to 1200 | Up to 1200 |
| Termination | 2-wire or 4-wire | 2-wire or 4-wire |
| Envelope Delay | 1000 - 2600 Hz 500 $\mu$s<br>600 - 2600 Hz 1500 $\mu$s<br>500 - 2800 Hz 3000 $\mu$s | 1000 - 2600 Hz 300 $\mu$s<br>800 - 2800 Hz 500 $\mu$s<br>600 - 3000 Hz 1500 $\mu$s<br>500 - 3000 Hz 3000 $\mu$s |
| Atten. with Freq. | 500 - 2800 Hz −1 to +3 dB<br>300 - 3000 Hz −2 to +6 dB | 500 - 3000 Hz −2 to +3 dB<br>300 - 3200 Hz −2 to +6 dB<br>(with 1000 Hz reference)<br>(+ means less loss) |

| | | |
|---|---|---|
| Common Carrier | AT&T | AT&T |
| Channel Desc. | Type 4003 (Sched. 1 Telephoto) | Type 4002 (Sched. 2 Telephoto) |
| FCC Tariff No. | 260 | 260 |
| Uses | Telephotograph transmission | Same as Sched. 1, but may be full duplex. |
| Band (+Bandwidth) | Approx. 1200 to 2600 | Approx. 1200 to 2600 |
| Operation Mode | Half-duplex | Half- or full-duplex |
| Termination | 2-wire or 4-wire | 2-wire or 4-wire |

| | | |
|---|---|---|
| Common Carrier | Bell Telephone Companies | AT&T |
| Channel Desc. | WATS (Wide Area Tel. Svc.) | Type 5500 (TELPAK A) |
| FCC Tariff No. | Local tariffs | 260 |
| Uses | | Voice, Data, or Alternate Voice and Data |
| Band (+Bandwidth) | Voice | Wideband (48 kHz) |
| Type of Service | Switched network | 2 - point or multipoint |
| Operation Mode | Half-duplex | Half- or full-duplex |
| Baud Rate | Up to 1000 | |
| Termination | 2-wire | Single Chan. or up to 12 Circuits |

| | | |
|---|---|---|
| Common Carrier | AT&T | AT&T |
| Channel Desc. | Type 5600 (TELPAK B) | Type 5700 (TELPAK C) |
| FCC Tariff No. | 260 | 260 |
| Uses | Voice, Data, or Alternate Voice and Data | Voice, Data, or Alternate Voice and Data |
| Band (+Bandwidth) | Wideband (96 kHz) | Wideband (240 kHz) |
| Type of Service | 2 - point or multipoint | 2 - point or multipoint |
| Operation Mode | Half- or full-duplex | Half- or full-duplex |
| Termination | Up to 24 Circuits | Single Chan. or up to 60 Circuits |

**TABLE 7-5. CHANNEL CHARACTERISTICS** (Cont.)

| | | |
|---|---|---|
| Common Carrier | AT&T | AT&T |
| Channel Desc. | Type 5800 (TELPAK D) | Type 7001 |
| FCC Tariff No. | 260 | 260 |
| Uses | | Video transmission (TV) |
| Band (+Bandwidth) | Wideband (1000 kHz) | Wideband |
| Type of Service | 2 - point or multipoint | |
| Operation Mode | Half- or full-duplex | |
| Termination | Single Chan. or up to 240 Circuits | |
| Common Carrier | AT&T | AT&T |
| Channel Desc. | Sched. A Program (Type 6005) | Sched. AA Program (Type 6007) |
| FCC Tariff No. | 260 | 260 |
| Uses | Radio Broadcasting | Radio Broadcasting |
| Band (+Bandwidth) | Wideband (5 kHz) | Wideband (8 kHz) |
| Baud Rate | Up to 1750 | Up to 3250 |
| Common Carrier | AT&T | Western Union |
| Channel Desc. | Sched. AAA Program (Type 6009) | Class A Data (60 WPM) |
| FCC Tariff No. | 260 | 260 |
| Uses | Radio Broadcasting | Low-speed Data |
| Band (+Bandwidth) | Wideband (15 kHz) | Subvoice |
| Type of Service | | 2-point or multipoint |
| Operation Mode | | Half- or full-duplex |
| Baud Rate | Up to 6000 | Up to 50 |
| Termination | | 2-wire or 4-wire |
| Common Carrier | Western Union | Western Union |
| Channel Desc. | Class B Data (75 WPM) | Class C Data (100 WPM) |
| FCC Tariff No. | 237 | 237 |
| Uses | Low-speed Data | Low-speed Data |
| Band (+Bandwidth) | Subvoice | Subvoice |
| Type of Service | 2-point or multipoint | 2-point or multipoint |
| Operation Mode | Half- or full-duplex | Half- or full-duplex |
| Baud Rate | Up to 57 | Up to 75 |
| Termination | 2-wire or 4-wire | 2-wire or 4-wire |
| Common Carrier | Western Union | |
| Channel Desc. | Class D Data | |
| Band (+Bandwidth) | Subvoice | |
| Type of Service | 2-point or multipoint | |
| Operation Mode | Half- or full-duplex | |
| Baud Rate | Up to 180 | |
| Termination | 2-wire or 4-wire | |
| Common Carrier | Western Union | Western Union |
| Channel Desc. | Class E Data | Class F Data |
| Band (+Bandwidth) | Voice | Voice |
| Type of Service | 2-point or multipoint | 2-point only |
| Operation Mode | Half- or full-duplex | Half- or full-duplex |
| Baud Rate | Up to 1200 | Up to 2400 |
| Termination | 2-wire or 4-wire | 2-wire or 4-wire |
| Envelope Delay | 200 μs between 1000 and 2400 Hz | 500 - 2800 Hz 1000 μs<br>600 - 2600 Hz 500 μs<br>1000 - 2600 Hz 165 μs |
| Atten. with Freq. | 300 - 999 Hz −0.8 to +2.0 dB<br>1000 - 2400 Hz −0.5 to +1.0 dB<br>2400 - 2700 Hz −0.8 to +2.0 dB | 300 - 500 −0.8 to +2.0 dB<br>500 - 2800 −0.5 to +1.0 dB<br>2800 - 3000 −0.8 to +2.0 dB |

**TABLE 7-5. CHANNEL CHARACTERISTICS (Cont.)**

| | | |
|---|---|---|
| Common Carrier | Western Union | Western Union |
| Channel Desc. | Broadband Sched. 1 (2 kHz) | Broadband Sched. 2 (4 kHz) |
| FCC Tariff No. | 246 | 246 |
| Uses | Low-speed Data (low-quality voice also possible) | Voice/Data (Digital or analog) |
| Band (+Bandwidth) | Subvoice (2 kHz) | Voice (4 kHz) |
| Type of Service | Switched | Switched |
| Operation Mode | Half- or full-duplex | Half- or full-duplex |
| Baud Rate | 1200 | 2400 |
| Termination | 4-wire | 4-wire |
| Common Carrier | Western Union | Western Union |
| Channel Desc. | Broadband (8 kHz) | Broadband (16 kHz) |
| Uses | Voice/Data (Digital or Analog) | Voice/Data (Digital or Analog) |
| Band (+Bandwidth) | Broadband (8 kHz) | Broadband (16 kHz) |
| Type of Service | Switched | Switched |
| Operation Mode | Half- or full-duplex | Half- or full-duplex |
| Termination | 4-wire | 4-wire |
| Common Carrier | Western Union | Western Union |
| Channel Desc. | Broadband (48 kHz) | Facsimile (180 RPM) |
| Uses | Voice/Data (Digital or Analog) | |
| Band (+Bandwidth) | Broadband (48 kHz) | Voice |
| Type of Service | Switched | 2-point or multipoint |
| Operation Mode | Half- or full-duplex | Half- or full-duplex |
| Baud Rate | | 14 sq. in./min. |
| Termination | 4-wire | |
| Common Carrier | Western Union | Western Union |
| Channel Desc. | Facsimile (360 rpm) | Voice |
| Band (+Bandwidth) | Voice | Voice |
| Type of Service | 2-point or multipoint | 2-point or multipoint |
| Operation Mode | Half- or full-duplex | Half- or full-duplex |
| Baud Rate | 23 sq. in./min. | Up to 1200 |
| Termination | | 2-wire or 4-wire |
| Common Carrier | ITT, RCA, WUI | |
| Channel Desc. | DATEL | |
| Uses | Data and high-speed telegraph | |
| Type of Service | 2-point | |
| Operation Mode | Half-duplex | |
| Baud Rate | 600, 1200, or up to 1200 | |
| Termination | 2-wire | |

## Problems

1. What are five classes of transmission media?
2. Name and describe the three types of communications circuits.
3. To what does the term narrowband refer? Give examples of narrowband facilities.
4. Define VH coordinate and describe its use.
5. Define broadband.
6. Discuss several factors which should be considered before choosing a communication service for a particular use.

*CHAPTER 8*

# *Switching and Network Concepts*

## INTRODUCTION

Computing is no longer an infant science. As computing techniques became more mature, and as its speeds permitted increasing volumes of data, it often became acutely obvious that the data to be processed was not always generated at, or even near, the computer site, nor were the ultimate users of the resulting information always standing by the "out" window to pick up the material. The economics of centralized computing were also a strong factor in locating processing facilities in places often far removed from input sources and output destinations.

The communication of data is not new, either; telephone and telegraph networks have been carrying digital signals for more than a century. Due to its wide distribution, ease of access, and relatively low cost, then, it was a natural step to look to the exploitation of these networks for higher-speed communications between computers and other business machines.

Most early data communication systems were point-to-point connections established specifically for each individual requirement, and speeds were generally limited to a few hundred bps. Centralization and sophistication of computers, however, emphasized the need for higher speeds and increased flexibility in establishing connections. Today, the spread of computer "utilities" and the shared use of computing equipment would be virtually impossible if it weren't for the high-speed capabilities of the switched telephone and telegraph networks.

Even though the networks provide the possibility for exchanging huge volumes of data economically, they also impose certain restrictions that should be considered in the study of teleprocessing systems. This chapter will discuss briefly some of the concepts and characteristics of switching systems. The switched telephone network will receive particular emphasis, since it provides the majority of the channels and switching used by today's data communications. Where differences exist between it and the telegraph network, these will be discussed.

## SWITCHING FUNDAMENTALS

### Reasons for Switching

It obviously would be impractical, if not impossible, to arrange a permanent telephone connection between all terminals that ever had occasion to be connected. If only 6 businessmen had mutual interests, for example, 15 lines would be required to allow any one of them to talk with any other one if no common switching system were used (see Fig. 8-1).

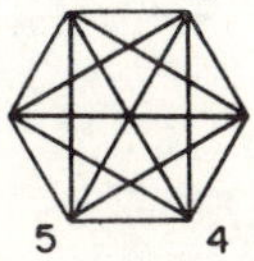


Figure 8-1. *Fifteen lines serve six customers and permit three separate and simultaneous converstations.*

Under these same conditions, a town with 10,000 people would require 50,005,000 separate lines (if everyone wanted the ability to call anyone else) and each customer would require exceedingly complicated and expensive equipment in his home.

To find the number of lines required to connect N stations, so that all are interconnected, use Gauss's formula:

$$\text{No. of lines} = (N-1) + (N-2) + \ldots (N-N)$$

The six customers would require

$$(6-1) + (6-2) + (6-3) + (6-4) + (6-5) + (6-6) = 5 + 4 + 3 + 2 + 1 + 0 = 15 \text{ lines}$$

Another way to the same solution is

$$\text{No. of lines} = \frac{N^2 - N}{2}$$

The group of six customers would thus require

$$\text{Lines} = \frac{N^2 - N}{2}$$

$$= \frac{6^2 - 6}{2} = \frac{36 - 6}{2} = 15$$

and each station added to such a system would cause the installation of N more lines. If a telephone switching center (central office; see Fig. 8-2) were installed in the town of 10,000, then only one line per customer, or 10,000 lines, would be required instead of about 50 million.

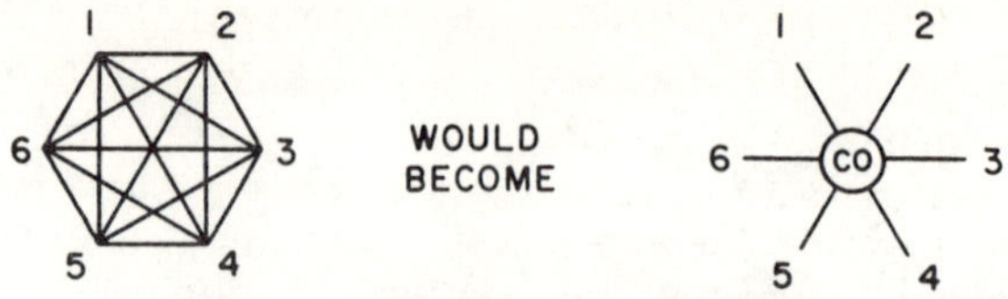


Figure 8-2. *Diagram of central office.*

The first reason for the use of switching, then, is for more efficient line use. The second is to permit alternate routing of calls, especially between cities and across the country. This subject is expanded later.

## Types of Switching Functions

Briefly, switching functions are classified in three categories: concentration, distribution, or expansion (see Fig. 8-3). An example of concentration switching is that done by the telephone switching systems found in most businesses. Calls to an outside number from any of the many telephones in the building are funneled out through one of the few switchboard trunks. The distribution, or routing, of these calls is done by the telephone company's switching central office, which acts under the instructions of the dialed number. Expansion is exemplified by the switchboard operator connecting the incoming trunks with particular office telephones.

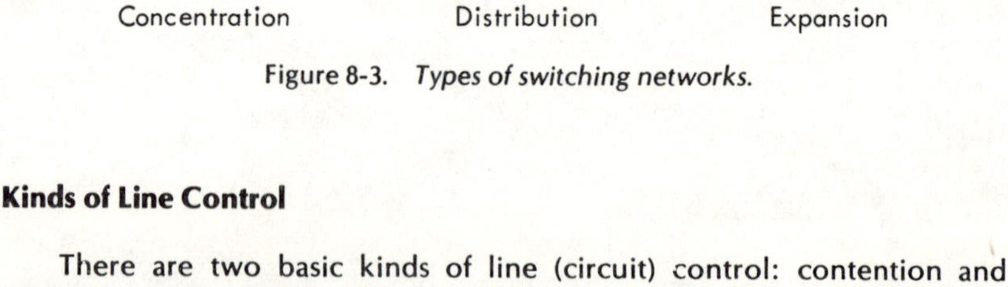


Figure 8-3. *Types of switching networks.*

## Kinds of Line Control

There are two basic kinds of line (circuit) control: contention and central. The term contention comes from several stations sharing a common circuit and each vying for it as it becomes available. A circuit with few stations, little traffic, or both, may cause few problems, but access becomes more difficult as volume increases.

To overcome station contention for the line, the ability to independently gain line access can be withheld from the individual stations, and control given to one station which can direct circuit traffic. This central control permits orderly and efficient use of the circuit by all stations.

### Circuit Modes

If a circuit is quiet (no activity), it is said to be in an *idle* mode. As the circuit becomes active in preparation for traffic transmittal, it is being used in the *control* mode. One of two conditions can exist in this mode: polling or addressing. *Polling* takes place when the control station sequentially calls each station on the circuit (including itself) to see if it has any traffic to send. If the called station does want to transmit traffic, control of the circuit is relinquished by the control station to the polled station. When the message has been transmitted, control is taken back by central control.

Facilities are usually provided at central control to change the polling pattern in several ways. Stations with heavier traffic can be polled more often than light stations. Stations also can be switched to manual polling or be removed from the polling sequence entirely. Since the master station often relays messages from other circuits, it can be picked up on every other poll, every third poll, or at any interval that prevents the accumulation of a backlog of outgoing messages. *Addressing* is the act of calling a station and indicating that it is to prepare to receive traffic.

The exchange of signals between machines during the setup period is called *handshaking*. This includes verification that the called machine is the particular one desired, establishment of one machine as master and the other as slave, and the determination of the direction of the initial transmission.

The *message* mode (or text mode) commences at the conclusion of the control mode and continues until the message has been transmitted. At this point the circuit would revert to either an idle or a polling mode.

### Message Types

A message intended for only one terminal on the circuit is called an individual message. A message sent to a particular group of terminals is a group message. A message sent to all terminals on a circuit or within a system is a broadcast message.

### Types of Networks

A communications circuit is a line or channel connecting terminals. A network consists of one or more circuits. The most simple network, then, consists of one circuit between two terminals and is called *point-to-point*.

If additional terminals are linked to the point-to-point network, it becomes a *multipoint* network. A network operating with no control station is called a contention network, as mentioned above. One with a control station, obviously, is a controlled network. These terms are generalizations, since there may be any number of variations and combinations of networks.

### Switching Methods

There are two basic switching methods, *line* and *message*. Line switching refers to the establishment of a direct connection between the originator and the recipient of a message. The message is sent directly with no intermediate handling, and the opportunity exists for acknowledgement or request for corrections or clarification.

Message switching involves the transmission of text from a terminal to a relay point where it is held, if necessary, until the circuit to the receiving station is available. Since no direct connection between principles is made, it is not necessary that the stations be free simultaneously before traffic can be moved.

In general, the advantages of line switching include: (1) Two-way communications between terminals; (2) assurance of message delivery; (3) relatively inexpensive switching equipment; (4) message receipt at time of transmission; and (5) immediate reply or information feedback.

The advantages of message switching are: (1) Less expensive station equipment; (2) speed and code translation possible; and (3) improved utilization of lines (messages are forwarded immediately upon access to a line rather than when both lines are available).

## THE SWITCHED NETWORKS

### Telephone Numbers

More than 200 million calls a day are made between the more than 50 million telephone numbers in the United States. Since every telephone can call any other telephone, about $1.25 \times 10^{15}$ (1,250,000,000,000,000) different combinations are possible (combinations $= (N^2 - N)/2$ where $N =$ number of numbers).

All-number calling (ANC), despite the furor raised against it, was installed for a good reason: the telephone companies were reaching the capacity of the letter-letter-number system of central office prefix codes such as GL7, AX3, and the like:

| | |
|---|---|
| 8 | possible combinations on first two |
| × 8 | letter pulls (on the telephone dial) |
| 64 | |
| − 4 | meaningless combinations |
| 60 | |
| × 9 | possible digits for third number of prefix |
| 540 | possible central office prefix codes |

All-number calling made available a maximum of 189 additional combinations for a total of 729 ($9^3$). (There are 10 digits around the dial, but only nine are used for central office codes.)

## Numbering Plan Areas

Even with all-number calling there are not enough telephone numbers to satisfy requirements. There are 10,000 possible combinations of the four numbers following each of the 729 central office prefixes, which yields 7,290,000 numbers. But over 50 million numbers are in use. This is made possible by the division of the country into numbering plan areas (NPA), each with its own area code.

Area codes are allocated according to state or provincial boundaries and the number of telephones (or central offices) involved. Area codes are of the following forms: N0X or N1X. (See Fig. 8-4.)

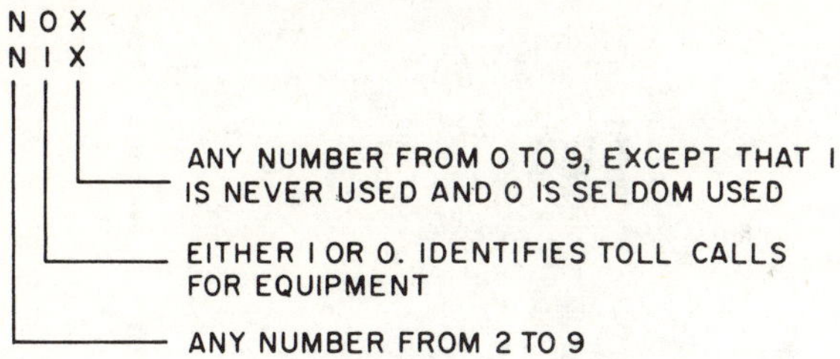


Figure 8-4. *Allocation of area codes.*

About 125 area codes are in use in the United States, Canada, and Mexico. About 25 remain for further assignment.

In areas that have been converted to ANC, a local call takes the form NXX-XXXX (such as 764-1234). In some areas, prepaid station-to-station calls are of the form 1+NXX NXX-XXXX. It is also possible to directly dial person-to-person and collect calls in some areas. The number is in the form 0+NXX NXX-XXXX, and the operator comes on line, after the number has been dialed, to get the required information about the call.

## Toll Switching Plan

All telephone numbers must adhere to an exact pattern because they must be recognized and interpreted by telephone switching equipment spread over an ever-increasing portion of the world. The toll switching plan in use in the United States provides reliability, economy, and a high probability that a user will both have access to the system and be routed to his destination.

The routing pattern of the toll switching plan is shown in Fig. 8-5. The solid lines represent final trunk groups, whereas the dotted lines are high-usage groups. High-usage trunks are always tried first. If they are busy, final trunks are tried. If these, too, are busy, the call is blocked.

Regional centers are located at: White Plains, N. Y.; Wayne, Pa.; Rockdale, Ga.; Pittsburgh, Pa.; Chicago, Ill.; St. Louis, Mo.; Dallas, Tex.; Denver, Colo.; Sacramento, Cal.; and San Bernardino, Cal.

Control switching points (CSP) are Class 1, 2, and 3 switching centers, and have these capabilities:

(1) Alternate routing — the ability to divert a call to an alternate trunk group if the first choice(s) is filled.
(2) Translation -- the ability to find the destination and *then* determine the routing.

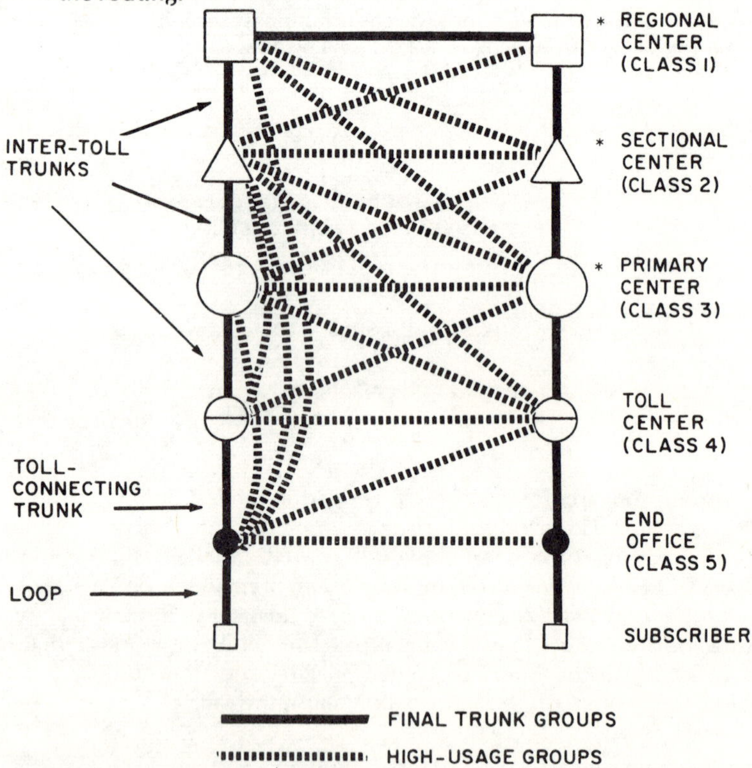


Figure 8-5. *Simplified switching plan diagram. Note: (1) These are control switching points (requirements listed in text). (2) Switching is always done toward the called station. (3) A maximum of two regional centers are involved in any call. There is no alternate routing on inter-RC trunks. (4) A maximum of seven intertoll trunks may be used between subscribers. (5) The toll center is a reasonable division between long-haul and short-haul systems. (6) High-usage trunks (dashed lines) may be established between any two centers if required by traffic volume.*

(3) Digit storage — done by the switching equipment.
(4) Code conversion — adding or subtracting digits in telephone number prefixes.

If all data could be sent over a single circuit between the two points involved, the characteristics of the circuit could be adjusted to the best operating conditions. A call placed over the switched message network, however, may involve as many as seven intertoll trunks plus four loops and intertoll connecting trunks. Because of alternate routing capabilities, a call placed to a point a thousand airline miles away may travel over several thousand miles of circuits before reaching its destination.

**Kinds of Switching Control**

There are two methods of switching control used in telephone central offices: progressive control and common control. In essence, progressive control responds to each turn of a telephone dial, and attempts to establish each leg of the connection as the number is dialed. If, at some point along the call's path, the circuit cannot be set up due to trunk groups being used to capacity, the call is blocked and the caller is given a busy signal. Progressive switching control lacks the capability of rerouting.

Common control switching overcomes this disadvantage by first checking a proposed connection from end to end and selecting appropriate alternate routes if necessary. To do this, it must absorb the entire number dialed by the calling station, determine the primary route the call should take, select any required alternatives, and test to see if the called number is busy. Only after the entire circuit has been analyzed and set up will the calling and called phones be connected and the common control drop from the circuit to perform other duties.

**Broadband Exchange Service (Western Union)**

Late in 1964, Western Union introduced the Automatic Broadband Switching System. It is designed to provide customers with an assortment of bandwidths wich may be switched through to other subscribers or customer locations, and paid for on a time-usage basis. Previously, only voice-grade service could be switched by the customer — all services wider than voice (such as TELPAK) were available on a full-time leased basis only.

The service is switched by using the seven-digit numbering system (sometimes prefixed by a three-digit foreign system area code) that appears in Table 8-1.

The initial Broadband Exchange Network consists of seven primary switching centers and 12 concentration locations which serve 22 cities (Fig. 8-6). Subscribers are given bandwidth ratings, and they may place calls using bandwidths at or below this rating (see Fig. 8-7).

## SWITCHING SYSTEMS

### Introduction

The switched telephone network is a well-used vehicle for medium-speed data transmission, and it is bound to become more popular as its potential is realized. It is important, therefore, to include a reference to the several types of network switching systems now in use. The brief descriptions below will provide a basic understanding of the operating fundamentals of each system and the effects each has on the transmission of data.

### Step-by-Step (Progressive Control)

The most common switching equipment found in telephone central offices is step-by-step (S×S). The name comes from the way the switches respond directly to the pulses of the calling telephone's dial. Fig. 8-8 is a simplified schematic of an S×S system. As the calling telephone is taken "off-hook," a line-finder seeks and connects to the calling line. Each line finder is connected back-to-back with a selector, which gives dial tone to the calling telephone. Under control of the pulses of the telephone dial, the selector contacts step up to the level of the number dialed, then automatically sweep across to the first idle line terminal. (In a telephone central office, this line terminal is connected to another bank of selectors, which, in turn, are connected to further selectors, outgoing interoffice trunks, and so on. A series of five selectors, one for each of the first five digits of the familiar seven-digit telephone number, are chained together — across town or across the country — on regular outside telephone calls. A call to another

**TABLE 8-1. BROADBAND NUMBERING SYSTEM**

| 10-digit Address → YYY | YYY | X | | XXX |
|---|---|---|---|---|
| Foreign System Area Code | Broadband Office Code | Bandwidth | | Called Subscriber Designation |
| | | Dial Number | Band | |
| YYY | YYY | 0 | Restricted | XXX |
| | | 1 | not used | |
| | | 2 | VF | |
| | | 3 | not used | |
| | | 4 | not used | |
| | | 5 | not used | |
| | | 6 | 2 kHz | |
| | | 7 | 8 kHz | |
| | | 8 | 16 kHz | |
| | | 9 | 48 kHz | |

Source: Western Union.
YYY – designated code, limited in use, by foreign systems
XXX – any digit, 0 to 9

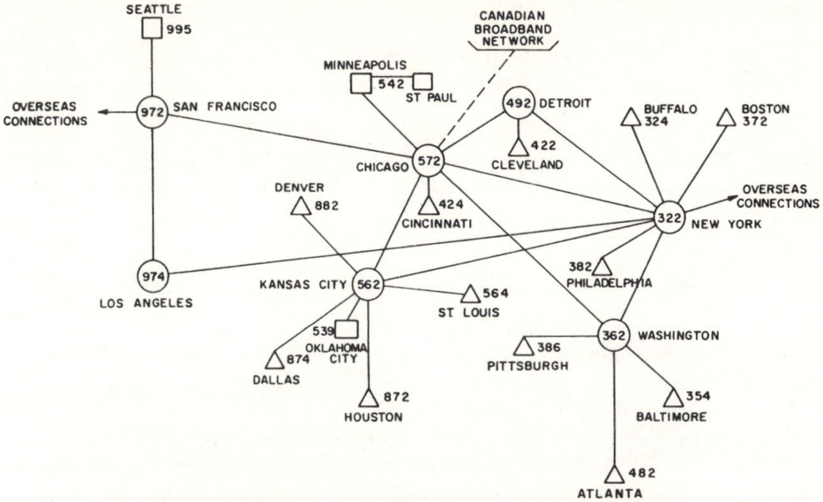


Figure 8-6. *Broadband exchange network. (Courtesy of Western Union)*

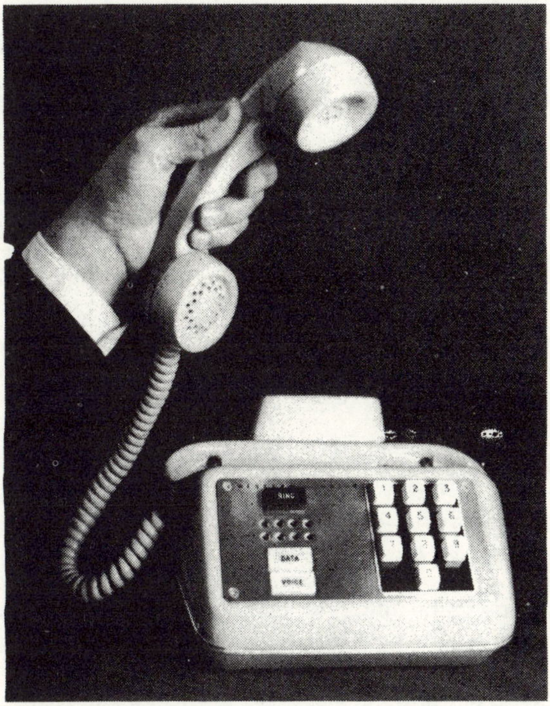


Figure 8-7. *Western Union voice-data instrument for use with broadband service exchange.*

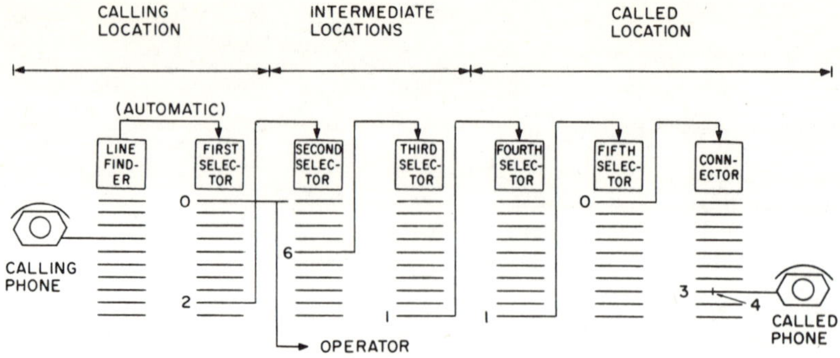


Figure 8-8. *Simplified schematic of a step-by-step switching system.*

telephone in the same building, served by a PBX, may use only one or two selectors.) The final selector connects the calling line to a connector associated with the called telephone's equipment. As the next-to-last number is dialed into the calling telephone, the connector steps up to the level of the number dialed. As the last number is dialed, the connector contacts sweep horizontally to the line corresponding to the number dialed. The connector then makes a busy test and either returns a busy tone to the calling telephone or rings the called telephone. All of the switches (line-finders, selectors, and connectors) remain in use until the calling party hangs up (calling-party hold), at which time they all drop off and become available for other calls.

The advantages of S×S equipment include its simplicity, economy, and complete modularity. Some disadvantages are that all equipment associated with a call is held for the entire length of the call; the call's route is under direct sequential control of the dialing telephone, prohibiting rerouting in case of a loaded trunk group; and the making and breaking of the switch contacts cause impulse noise on lines in close proximity to the switches. This is why S×S central offices are known as noisy offices. The switch connections in Fig. 8-8 were caused by the calling telephone dialing 261-1034.

## Crossbar (Common Control)

There are two main differences in the operation of a crossbar switching system as compared to the S×S system: (1) the dial pulses do not directly activate a series of stepping switches, and (2) the common equipment is occupied only during the initial and final stages of any individual call.

Whereas the S×S system was a slave to the dialed pulses from the sending station, the crossbar system absorbs the entire number, analyzes the steps necessary to perform the connection, seizes the necessary equipment and lines, tests the circuit, connects the end stations, and drops off.

The crossbar switch is basically a grid with vertical and horizontal

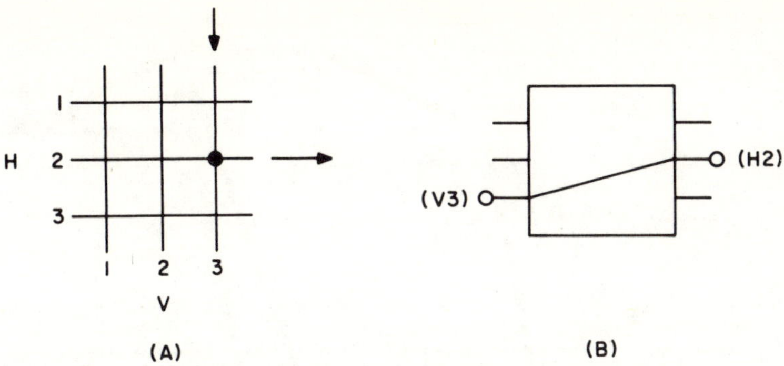


Figure 8-9. *(A) Simplified diagram of crossbar switch; (B) switch-block operation.*

movable bars. In the simplified diagram of Fig. 8-9(A), a connection has been made between vertical 3 (V3) and horizontal 2 (H2) (by activating the V3 and H2 relays). This same operation may be shown by the switch block in Fig. 8-9(B):

If several switch blocks are associated, alternate paths from input to output can be established (Fig. 8-10).

In reality, crossbar switches have hundreds of contacts (points), depending on the modularity. The modules are called *frames* (Fig. 8-11). Line link frames (LLF) are connected to subscriber's lines or interoffice trunks, and act as concentrators; they link from 190 to 590 lines to 100 switching trunks. Trunk link frames (TLF) are placed between groups of LLF and perform the actual switching function between incoming and outgoing lines under control of the marker. The marker (MKR) is the brains of the crossbar switcher. It interprets the dialed digits, determines the primary and alternate routing, locates and seizes idle trunks, relays outpulsing information, and establishes the connection. It performs its functions in about 1/10 of a second per call. The marker is assisted in its work by an originating register (OR), the main function of which is the interpretation of dialed and other information (such as class of service) for the marker.

The steps to establishing a crossbar connection, in oversimplified form, are:

(1) The marker is signaled by the calling telephone going off hook.
(2) The marker instructs the TLF to connect to the OR, then the marker releases.
(3) The TLF passes the calling line (telephone) number, the called number, and other information (first or second party, coin service, etc.) to the OR, which in turn passes it to the marker.
(4) The marker determines all connections and tests for busy.
(5) The marker makes the connection, applies ringing, and drops off.
(6) Called party answers, ringing drops off, and a through connection is established.

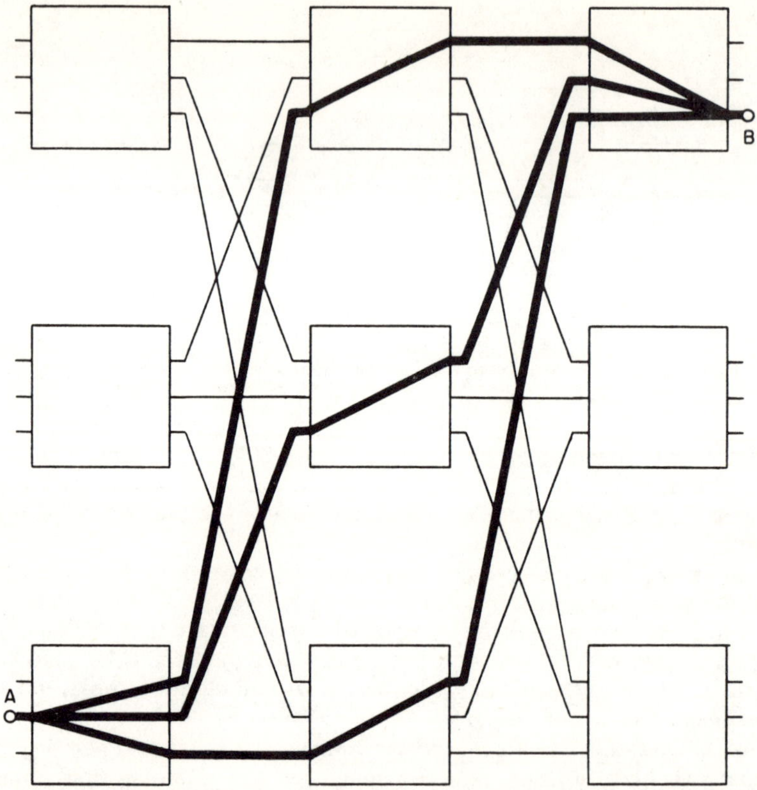


Figure 8-10. *Simplified crossbar switching matrix. Heavy lines show alternate routes between points A and B.*

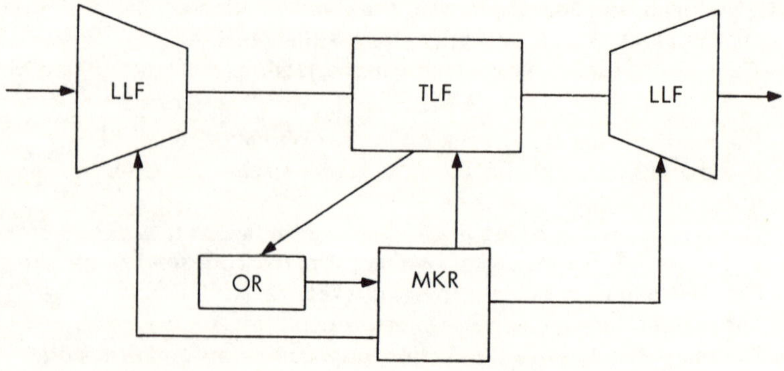


Figure 8-11. *Relationship of crossbar switching modules.*

The advantages of a crossbar switching office are the lower cost and much lower maintenance as compared to S×S, alternate routing capabilities, and especially important to data communications, noise caused by switch contacts is reduced dramatically. Dialing time is also shortened considerably; this is particularly important if a computer is being used to poll outlying stations. The average time required to dial 10 digits by each of the four methods is:

| Method | Seconds |
|---|---|
| Rotary dial pulsing (10 pulses/sec) | 17.0 |
| Repertory pulse dialer (10 pps) | 8.5 |
| Manual Touch-tone | 3.5 |
| Repertory Touch-tone | 1.1 |

## Electronic (No. 1 ESS)

Although crossbar switching in many ways was a great improvement over step-by-step systems, its speed was still limited by its electromechanical components. An electronic switching system (No. 1 ESS) developed by the Bell Telephone System overcomes this speed limitation and also provides a variety of new services.

Like crossbar switching, the ESS is a common control space-division system. Due to the much faster electronic circuitry, however, the central control not only can handle a great number of lines, but can also perform administrative and maintenance functions.

A simplified block diagram of the No. 1 ESS system appears in Fig. 8-12 and its operational details are shown in Fig. 8-13. The sequence of operations is:

(1) Caller goes off-hook, causing current to be placed on his line and a ferrod sensor to change state.
(2) During its normal cycle, the Control Unit compares the sensor's condition with its previous condition. The change of state causes a Digit Detector to give dial tone to the caller.
(3) The Digit Detector is scanned every 10 ms to detect dial pulses. When the entire telephone number has been recorded, it is given to the Program Store, which determines the nature of the call.
(4) Memory is checked for the state of the called line, the state of the calling line is changed to busy, and line links are reserved for a talking path.
(5) Ringing current is applied to the called line, ringing tone is given to the caller.
(6) Called party answers, ringing current and tones are stopped, and the connection is made.
(7) When the call is terminated the connection is broken and all associated memory registers are restored.

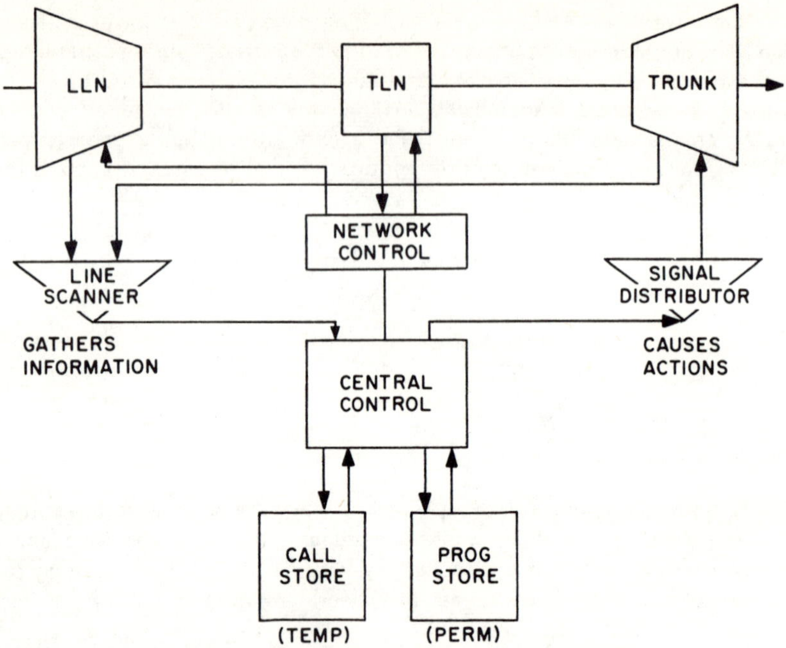


Figure 8-12. *No. 1 electric switching system (ESS).*

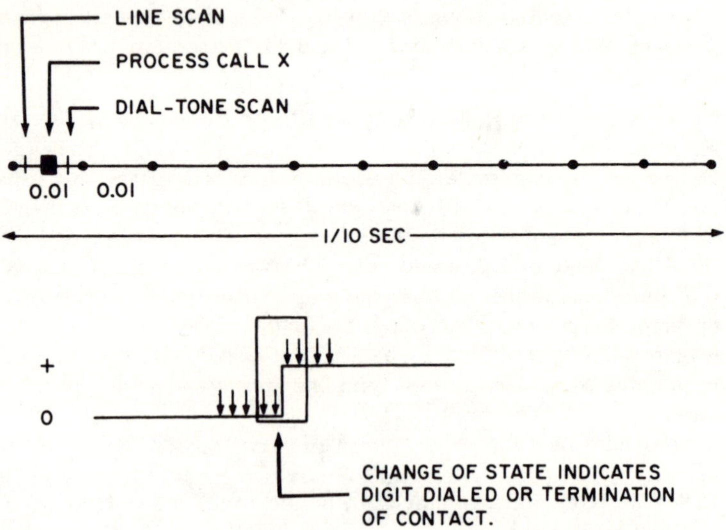


Figure 8-13. *Details of No. 1 ESS operations.*

Advantages of ESS include greater line capacity, a minimum of error-causing disturbances, greatly increased dependability, better economy, a self-maintenance program, and an increased variety of customer services. All of these are important in a teleprocessing system.

**Time-Division (No. 101 ESS)**

All the systems discussed to this point have been space-division switching systems; every connection required an individual line through the switching center (PBX or central office). The No. 101 ESS introduces the concept of time-division switching.

Figure 8-14 gives a graphic comparison of space-division and time-division switching. Although both have a central control, the number of connections between stations in the space-division system are limited to the number of physical lines (wire pairs) between station groups. The time-division example carries many simultaneous connections over its one line (bus), the number being limited only by the internal design of the equipment.

Rather than rapidly sampling and acting upon the condition of many lines, as does the No. 1 ESS, the No. 101 samples segments of time that have been allotted to the stations in the system. These segments are called *time slots,* and up to 25 time slots are associated with a common line, or bus. Under control of a central unit, the time slots are sampled for 1 μsec (microsecond) every 80 μsec, one at a time. The control unit assigns gates to the time slots to complete a connection as in Figure 8-15. Gates may be assigned to time slots which are associated with telephone sets, ringing current, audible rings, central office trunks, switchboard lines, and the like. Any gate may be assigned to any time slot to fit particular connection requirements.

In Fig. 8-15, the calling station has already dialed another station in the same building. The gates to the calling station and Audible Ring are operated simultaneously during a particular time slot, and the calling party hears the ringing tone. Within some other time slot, gates to Ringing Current and the called station are simultaneously operated, causing the called telephone to ring. When the called station answers, the ringing current is stopped, both stations are gated to the same time slot, and conversation can begin. To preserve the fidelity of any signals we may introduce to the system, sampling is done at twice the frequency of the highest frequency desired for transmission. The 1 μsec sample taken every 80 μsec produces a signal of 12,500 Hz.

The call sequence of the No. 101 ESS is as follows (see Fig. 8-16):

(1) An off-hook condition is detected by the scan, and the Input-Output is notified in binary over the data link.
(2) Call Status Store is given the identification of the calling station. The Digit Receiver, time slot, and Digit Trunks are assigned by Program Store, which connects dial tone to Digit Trunks.

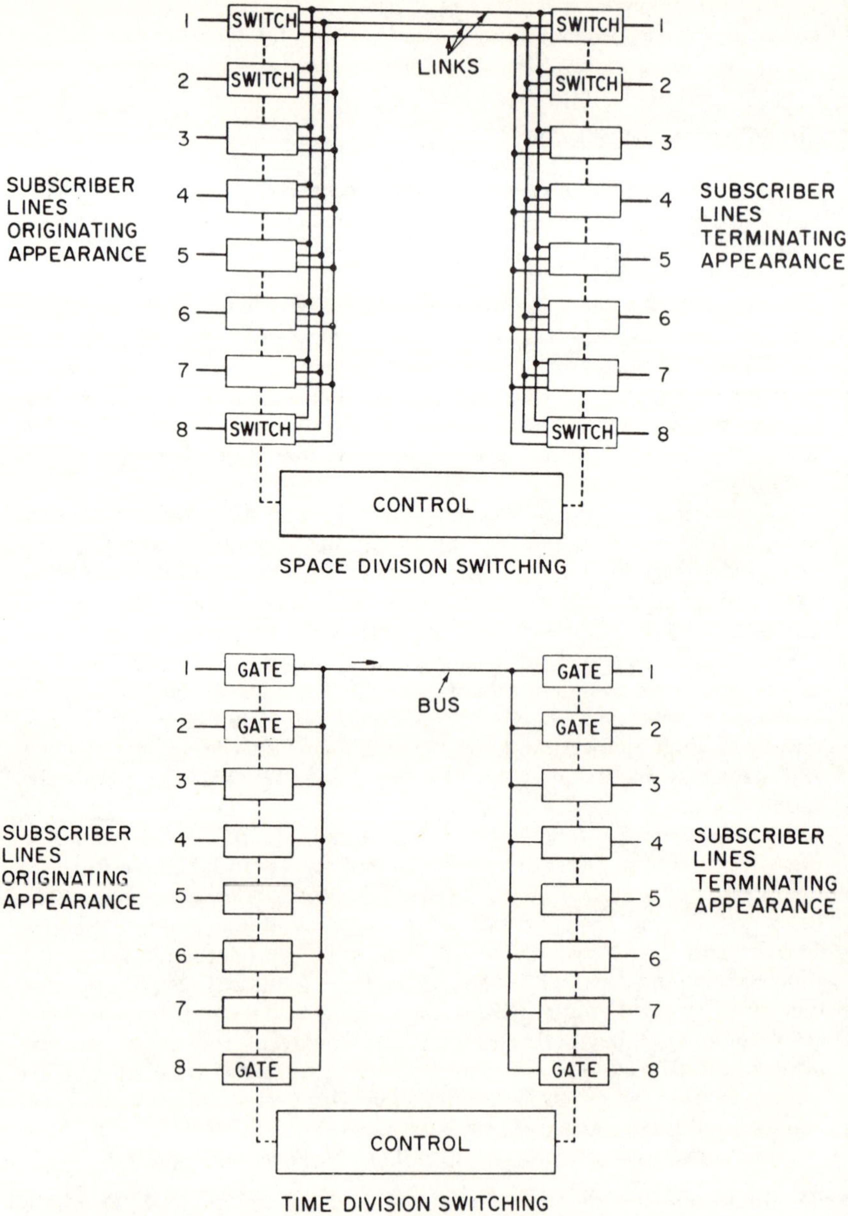


Figure 8-14. *Comparison of space-division and time-division schematics.*

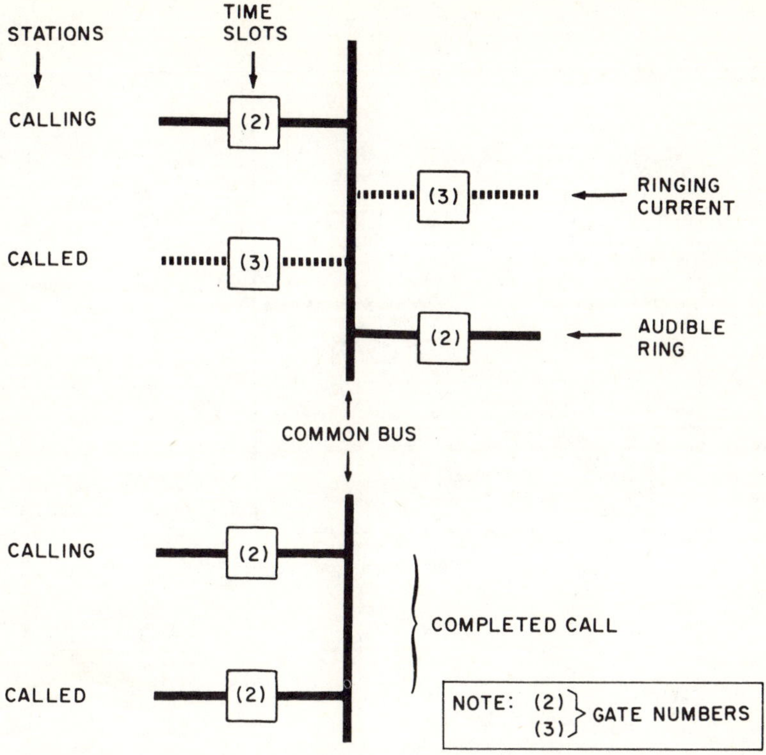


Figure 8-15. *Steps in establishing a connection using time-division switching.*

(3) Dial tone is given to the calling station by the Digit Trunks and calling line gates being operated simultaneously.
(4) The multifrequency tones from the calling stations dial are forwarded by Local Control to the Digit Receiver, which sends the accumulated number to Call Status Store.
(5) Ringing current is assigned a time slot with the called station. When the station answers, ringing current is stopped and both stations are gated to the same time slot to complete the connection.

The No. 1 ESS (for use in central offices) and the No. 101 ESS (for customer use) are both operational today. The Bell System does not intend to replace its present $16 billion switching equipment inventory overnight, but, as with the change from manual to dial calling, ESS equipment will gradually become a greater proportion of operating systems. Present guesses put complete ESS coverage at about the year 2000.

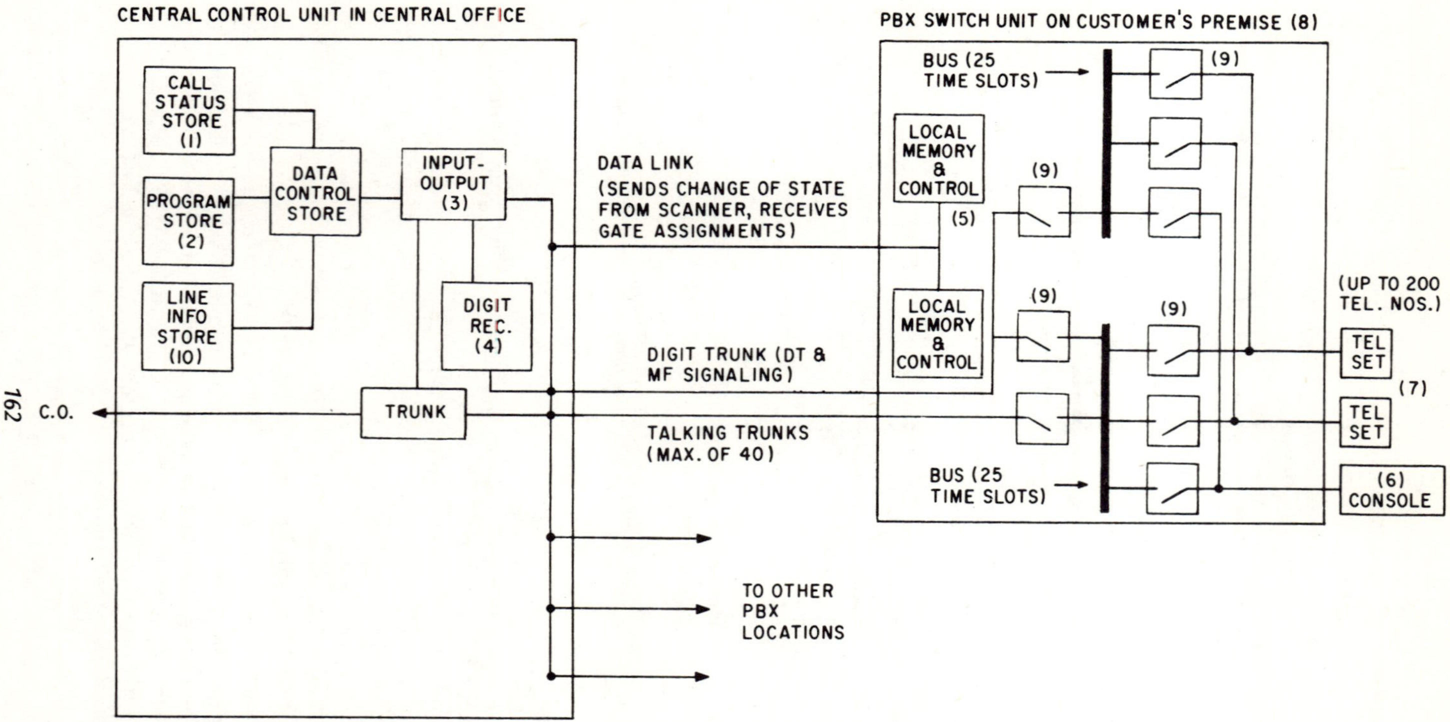


Figure 8-16. *No. 101 ESS schematic. Note: (1) Ferrite memory (read and write). Contains status of all calls in progress. Is scanned periodically. (2) Twistor memory (read only). Main program. (3) Input: changes of state and dialed digits. Output: switching instructions and outpulsing digits. (4) Gives dial tone. Receives multifrequency (NMF) digits. (5) Takes a 1 μs sample of each time slot every 80 μs. (6) Always MF (multifrequency, or tone dialing). (7) MF (multifrequency) or DP (dial pulse). (8) All scanning and time-division done here. (9) Gatews which operate under direction of local Memory and Control (see note 5). (10) Contains class of service and line location.*

**Noncomputer Systems (Private-line Teletypewriter)**

Communications requirements vary so much between industries, and even between companies within industries, that a generalized switching package designed to answer all communications problems is nothing but an utopian dream. About two-thirds of the switching plans offered by Western Union, for example, were designed for a special customer or purpose. Although there are several versions of private-line teletypewriter switching systems available, most of them closely resemble the composite system described below.

Although the particulars may differ, the fundamentals are common to most switching systems. A composite example consists of one or more lines, or circuits, each one of which connects up to 20 stations. The station may consist of teletypewriters arranged for automatic send and receive (ASR), keyboard send-receive (KSR), or receive only (RO). Peripheral equipment for tape handling may also be used.

Control equipment causes each station on the line to be polled automatically. Polling consists of sending two-letter transmitter start codes (TSC) on the line. All stations "listen" to all TSC's, but respond only to their own. The polling sequence includes the master station. If a polled station (either master or slave) has no messages to transmit, its transmitter responds with a letter V answerback. If there is a prepared tape in the transmitter of a polled station, the transmitter will turn on automatically. (See Table 8-2.)

The first part of every message contains a two-letter call-directing code (CDC). All stations on a line listen to all CDC's, but a receiver will turn on only in response to its own CDC. Prearranged groups of stations can be turned on by group CDC's, whereas broadcast CDC's cause all stations to be activated. Called stations must answer with the letter V before the transmitting station can begin sending.

If the system consists of only one line, every station can send directly to any other station, and the primary duty of the master station is to control the sequence of station operations. If the system comprises more than one line, there must be some way to transfer messages from one line to another. Since the receiver's line may not be available at the time the transmitter begins to send, there must also be some way to store the message and forward it when the receiver is clear.

Both the transferring and the storage functions are accomplished by capturing the message in punched paper tape. The tape is prepared by a reperforator at the control point, the portion containing the message is torn off, and the tape is fed into a tape transmitter (reader) on the receiver's line. When the line becomes free, the tape is transmitted. (A reperforator is a paper tape punch that prepares a tape copy of a message as it is received from a remote transmitter. A tape perforator is a paper tape punch that receives its signals from an associated keyboard as the message is being prepared for the original transmission.)

**TABLE 8-2. TELEGRAPH LINE CONTROL SEQUENCES**

| Signal Type | AT&T<br>83B2<br>HDX | WU<br>115<br>HDX | AT&T<br>FINAC<br>FDX | WU<br>117<br>FDX | AT&T<br>81D1<br>FDX | WU<br>57<br>FDX | WU<br>111<br>Conten. |
|---|---|---|---|---|---|---|---|
| **Station Request to Send** | 1 | 1 | 1 | 1 | 1 | Space<br>or 1 | 3.0 Sec<br>Open |
| Polling Sequence<br>Initial<br>(from central) | EOM | EOM | 3 | FIGS<br>Z | Blank<br>0.4 sec<br>Pause<br>Space | Y | 2 |
| Transmitter Start Code (from central) | ALPHA<br>M or G | X<br>ALPHA | CR 3<br>ALPHA | ALPHA | ALPHA | ALPHA | 2 |
| Positive Response (from station) | Msg | Msg | Msg | Msg | Msg | Y | 2 |
| Negative Response (from station) | V or 2 sec | V or 10 sec | 1.0 sec | V or 2 sec | H or 5 sec | Space | 2 |
| End of Polling (from central) | | | 3 | 4 | 2 LTRS<br>0.2 sec<br>Pause | CR | 2 |
| Selection Sequence<br>Initial SOM<br>(from sender) | LTRS<br>LTRS<br>LTRS | LTRS<br>LTRS<br>LTRS | LTRS<br>LTRS<br>LTRS | FIGS<br>H<br>LTRS<br>7 | FIGS<br>H<br>LTRS<br>(optional) | W | LTRS |
| Selection Code (from sender) | ALPHA<br>ALPHA<br>LTRS | ALPHA<br>ALPHA | (ALPHA)<br>(option)<br>ALPHA<br>LTRS | Q 6<br>3 LTRS<br>2 ALPH | ALPHA<br>ALPHA<br>LTRS | ALPHA<br>ALPHA<br>5 | ALPHA |
| Positive Response (from addressee) | V | V | None | V 6 | None | ALPHA<br>(complement of Selection) | None |
| Negative Response (from addressee) | 10 Sec | ? | None | ? 6 | None | ? | None |
| End of Selection (from sender) | CR LF | Space<br>CR LF | CR | Space | CR LF<br>LTRS | Space<br>Space | CR LF<br>Space<br>LTRS |
| FOM | FIGS<br>H LTRS | FIGS<br>H LTRS | FIGS<br>H LTRS | FIGS<br>H LTRS | FIGS<br>H LTRS | FIGS<br>H LTRS | Not<br>Fixed |
| EOT | | | | | EOM<br>+<br>H LTRS | | |

Source: IBM

General Notes

- Polling sequences normally include several TSC codes, each followed by a response, except as noted.
- The term ALPHA refers to any alphabetic character from the specific list for the particular system.
- In HDX systems, a message must start with a selection sequence, even if no selections are included.
- In FDX systems, a message from a station can only be sent to a central or control station. Messages to stations are only sent from the central or control station.

Specific Notes (as keyed to table):

1. Station request recognized only as the result of a poll.
2. This is a contention system – each station controls its own entry to the line using a Way Station Selector 7304.
3. A TSC is sent following every CR in the text. If there is traffic on the incoming line (no poll required), a BLANK is sent instead of the ALPHA. Terminals on this system automatically initiate a CR and LF when they receive a CR.
4. The TSC code unblinds all previously selected printers for continuing outgoing traffic; thus, each TSC must be preceded by FIGS Z, but no end-of-polling code is needed.
5. Each WU57 Selection Code is always one alphabetic character repeated once.
6. Response to Selection is optional; a Q and three LTRS preceding the Selection Code will trigger the answer. Only the last Selection on each message can be required to answer back, and only one of a group code.
7. Must precede the first message in each transmission.

## Comparison of Line and Message Switching Functions

As we have seen from the above discussions, switching systems can be classified as being one of two categories: line switching or message switching. Briefly, a line switcher receives routing instructions and builds an end-to-end circuit path before accepting the message. The entire circuit is maintained while the message is sent from the originator directly to the recipient. A message switcher, on the other hand, accepts and stores entire messages from originators before establishing a forwarding circuit to the recipient. In fact, several cycles of store-and-forward may occur before a message is delivered to its addressee.

A computer line switcher can make very efficient use of a single central control (logic) unit, and its memory, for the most part, comprises the stored program and many small registers. The line switcher's main function is to monitor incoming and outgoing lines and respond to conditions found on them (telephone going off-hook, dialing pulses, call termination, etc.). Since central control merely establishes the routing and is not involved with handling the messages themselves, it is able to supervise the flow of a large quantity of information.

A message switcher, on the other hand, not only has to monitor the lines, but it must also absorb and store messages for future delivery and perform many other tasks, such as message numbering, traffic accounting, priority recognition, error checking, etc. Unlike the line switcher, which does one job at a time (a serial system), the message switcher must be performing many various operations simultaneously (a parallel system). It must have a large memory (or several memory modules) to store incoming messages, and adequate buffering between this memory and the lines or input-output (I/O) devices.

A good example of line-switching by computer is the telephone company's switched telephone network. It can be considered as one huge computer or, perhaps more realistically, as many compatible interconnected computers. Familiar examples of computer message switching are on a smaller scale, such as the IBM 7750 or the GE Datanet 30.

Line switchers are used to satisfy requirements for voice contact, immediate line access, and end-to-end communications. Message switchers nicely answer needs for transmitting written messages, where slower speeds and slight delays are not objectionable.

## Message Switching Functions

Listed below are some features which should be considered for inclusion in message switching systems (not necessarily in order of importance).

(1) Message storage and forwarding
   (a) Capability to handle single-address, multiaddress, group address, and broadcast messages

- (b) Multiaddress messages sent to individual stations as their circuits become available
- (c) Direct on-line transmission between stations on the same circuit, and with message accounting
- (d) Message queuing
- (e) Speed and code translation
- (f) Allowance for mix of terminal types
- (g) Multiple priority
- (h) Unattended message transmission and receipt at switcher and terminals
- (i) Variable polling — preferential treatment to stations with heavy traffic, restricted polling if system capacity is approached

(2) Message protection
- (a) Message saved if system fails
- (b) Checkpoint and restart procedures
- (c) Automatic self-diagnostic machine routines
- (d) Error control
- (e) Require acknowledgment of all incoming and outgoing messages
- (f) Message identification — automatic insertion of date, time, circuit number, and message serial number

(3) Message accounting
- (a) Proration of message and system costs to users
- (b) Statistical traffic and line-load analysis
- (c) Process message cancellations
- (d) Create reports which result from any analysis performed

(4) Message retrieval
- (a) Long-term message filing (hours or days)
- (b) Retrieval possible only from control console

(5) Message monitoring
- (a) Comparison of machine and mnemonic address codes
- (b) Format accuracy check
- (c) Printout of all abnormalities
- (d) Wilfull intercept of messages at central console
- (e) Automatic advice to stations using incorrect codes and/or procedures
- (f) Message editing (addition or deletion of information)
- (g) Error control

(6) Special functions
- (a) On-line link to central computer (for inquiry and information systems)
- (b) Ability to expand with growth
- (c) Programming flexibility to allow for changes
- (d) Initilization and shutdown (good morning and good evening) routines
- (e) Dynamic analysis of internal storage availability
- (f) Maintenance of a running analysis of line and terminal performance, with periodic printed reports

## Problems

1. What are the two primary advantages of switching systems?
2. Discuss the three functions of switching.
3. What are the two kinds of line control?
4. Describe the three circuit modes.
5. Compare polling and addressing.
6. Give the two methods of switching, and discuss advantages of each.
7. Draw a simplified switching-plan diagram, and label its parts.
8. Name and describe the two kinds of switching control.
9. Draw a crossbar switch grid, and indicate a connection between V4 and H2.
10. Give two advantages of crossbar switching over step-by-step.
11. Crossbar is a common-control, space-division switching system. What does this mean?
12. What is the prime advantage of time-division switching over switching done by space-division?
13. Define: ASR, CDC, RO, TSC.
14. Compare the functions of message switchers and line switchers.

CHAPTER 9

# *Data-Set Uses and Characteristics*

## DATA-SET PARAMETERS

### Introduction

In general, data sets perform the function of conditioning signals received from transmitting business machines so that the signals are acceptable to the transmission medium. At the receiving end, the conditioned signals are reconverted by the data set, and are presented in their original form to the receiving business machine. It is important to note that data sets do not originate data themselves, nor do they have any means to detect or correct errors. These are functions of the business machines. Some data sets (particularly the analog type) do not modulate the signal, but merely provide filtering and limiting functions to protect the business machine and the transmission medium. The following discussion of the various parameters should help clarify how data sets operate and why they are needed. (Data-set characteristics are charted at the end of this chapter.)

### Digital or Analog

A digital data set recognizes a limited set of discrete values (usually two), while an analog set recognizes a *range* of values between limits. Digital sets are used to transmit binary codes such as baudot and ASCII, while handwriting and facsimile are examples of signals that are not readily digitized, and thus require an analog data set.

### Coding (Binary, Ternary, etc.)

If the signals of a data set can assume either of two possible states, the data set is said to be binary. If the signal could assume any one of three

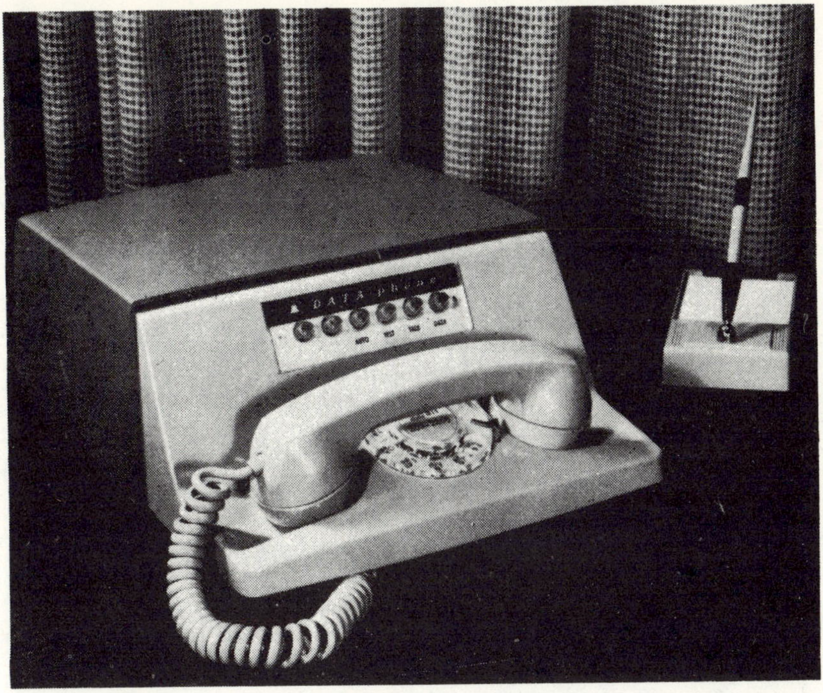

Figure 9-1. *A typical Data-phone data set. (Courtesy of AT & T)*

possible states, it would be ternary and so forth. This definition applies only to each channel of the data set's signals. A parallel set may have several channels, but the signals within any one of the channels is binary (or ternary, etc.)

## Serial or Parallel

A serial data set handles only one bit at a time. It is insensitive to the bits per character and to the coding scheme. A parallel set, on the other hand, accepts and transmits an entire character at once. There are several types of parallel data sets. One type uses two channels to send numeric data only. Another forms three channels and is capable of sending alphanumeric data. Both of these require signals that have been precoded by the business machine. A third type of parallel set forms eight channels, and is thus capable of handling any code up to eight bits per character. This type does not require precoding by the business machine and is code insensitive. (See Fig. 9-2.)

In comparison to serial data sets, parallel sets are slow. This is largely due to the channelizing of the voice band; much of the available bandwidth is absorbed as buffering between the data channels.

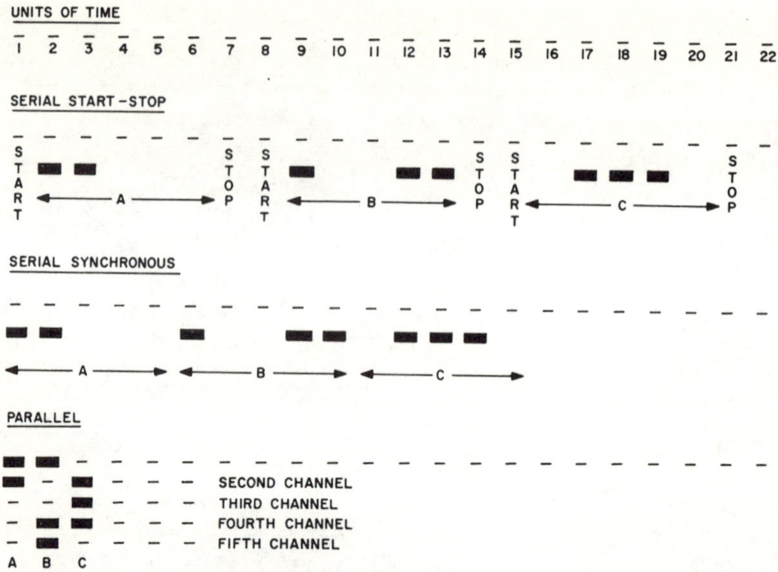


Figure 9-2. *Comparison of serial and parallel transmissions. (Courtesy of IBM)*

## Transmitter, Receiver, or Transceiver

Most of the serial data sets can transmit and receive, and many can do these simultaneously. The parallel sets and some of the analog sets, however, are designed to transmit only or receive only. The transmitting sets modulate the signal from the business machine, whereas the receiving sets demodulate the signal. Sets that both modulate and demodulate are often called "modems." Therefore, all modems are data sets, while all data sets are not modems.

## Simplex, Half-duplex, and Full-duplex

Data sets that are designed to either transmit or receive, but not both, are simplex sets (one way only). Those that can transmit and receive, but only one at a time, are half-duplex sets, and those that can transmit and receive simultaneously are called full-duplex sets.

## Speed

The speed of a data set is measured in bits, bauds, or characters per second. Serial sets are usually rated in bps, but since some send more than one bit per baud, it is important to know both values (bits and bauds). Parallel sets are normally rated in cps. If the number of bits per character are known (such as five for baudot, eight for ASCII, etc.), the corresponding number of bps can be calculated.

### Modulation

Amplitude, frequency, and phase modulation techniques are used by various data set to condition the business machine's signal for transmission. A few data sets do no modulation, but merely filter and limit the output of the business machine.

### Synchronous or Asynchronous

A *synchronous* data set accepts a serial stream of bits from a business machine at exact time intervals. Either the data set or the business machine contains an electronic clock that gates the bits to the data set (and to the business machine at the receiving end). These sets are rated *at* a certain baud (or bit) rate, usually ±.01 percent. If the data set provides the clock, it is said to have internal timing. If the associated business machine provides the clock, the data set has external timing. Be careful to specify your reference (the data set or the business machine) when speaking of internal or external timing. Data sets that accept an input signal at rates that vary between minimum and maximum speeds are called *asynchronous* sets.

### Attendance

It is possible to arrange many data sets so that when they are called from a distant point they will respond by turning on the associated business machine and begin operation, all without human assistance. This is called *unattended* operation. This is useful, for example, when information that has been stored in paper or magnetic tape or punched cards is placed in a transmitter at a remote location before the operator goes home at night. During the night the machine is called and automatically transmits the information with which it has been loaded. Information can also be transmitted to an unattended receiver at some remote point. An *attended* arrangement requires human assistance to perform transmission or receipt of data.

### Voice

Not all data sets have the capability to alternately send voice and data. Those with voice capabilities have an associated telephone to provide voice coordination between operators and to provide a means to dial the call. Sets used with unattended service or with automatic dialers do not always have telephones. The Bell Telephone System uses the trademark "Data-Phone" to designate those data sets which have alternate voice capabilities and thus can be used with the switched telephone network.

### Message Network or Private Line

The present switched message (telephone) network has certain characteristics that dictate the ways in which it can be used for data transmission. For example, data sets must not produce a signal of more than 1200 bauds if errors are to be kept to a minimum. The network will not accept d-c signals, so data sets are required to modulate such signals from business machines. The qualities of every line of the vast telephone network cannot be maintained to exact values, so data sets must be designed for less than peak performance.

A private line, however, is used only by the party who leased it. Its parameters, therefore, are known and can be kept at optimum levels. Because of this, data sets used over private (leased) lines can be geared to higher transmission speeds, and often do not require voice or dialing capabilities.

### Narrow, Voice, or Wideband

The speed of a data set (in bauds or characters) is geared to the type of facility over which it will be communicating. Sets designed to work at teletypewriter speeds, for example, (up to 75 bauds) will operate over very narrow bandwidths. Others operate up to the 1200-baud maximum of the switched telephone network, whereas still others operate in the wideband range of 40.8 kilobits and up.

### Classification of Bell System Data Sets

The Bell Telephone System has classified its data sets, by their general characteristics, into a series. The number variations within the series designates subseries, and letters after the numbers indicate variations. An explanation of the series numbers is given below, and characteristics of each data set are found in Figs. 9-5 to 9-7 on pages 176 to 178.

| Series | Characteristics |
|---|---|
| 100 | Low-speed serial (up to 300 bauds per second). |
| 200 | Medium-speed serial (between 300 and 1800 bauds). |
| 300 | High-speed serial (wideband). |
| 400 | Low- to medium-speed parallel (up to 75 cps). |
| 500 | High-speed parallel (wideband). |
| 600 | Voiceband analog. |
| 700 | Wideband analog. |
| 800 | Auxiliary data sets (line control, automatic dialing, etc.). |
| 900 | Test sets. |

### Reverse Channel

Many data sets coming on the market now have reverse channels which may be used for control, timing, or error-detection purposes. The reverse channel usually consists of a single-frequency tone sent back to the transmitter by the receiver. The business machine transmitter can be arranged to stop if the tone stops and to start or back up and retransmit if the tone reappears. Thus the transmitter can be controlled by the receiver, and the transmitter also would halt if the communication line were broken. Timing signals can be sent from one end to the other to synchronize operations. If some form of error detection and correction is used, the transmitter can continue to transmit until an error causes the receiver to request a retransmission.

## The Interface

Data sets are arranged to accept and deliver one of three types of interface conditions: contact closures, voltages, or currents. Most serial sets work with voltage or current interfaces, while the parallel sets usually use contact closures. The Electronic Industries Association (EIA) standard voltage interface is usually provided to facilitate connections between data sets and associated business machines. The following interface circuit definitions are taken from the EIA RS-232-B Standard and Bell System Technical Reference Manuals, and are keyed to the circuit names found on the diagram in Fig. 9-3. See also Fig. 9-4.

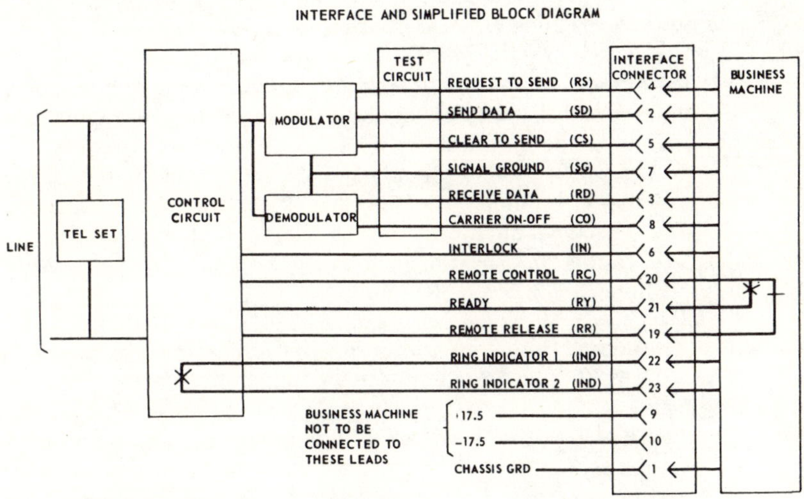


1. Frame ground
2. Send data
3. Receive data
4. Send request
5. Clear-to-send
6. Interlock
7. Signal ground
8. Carrier on-off
9. + Power
10. − Power
11. Not used
12. Not used
13. Not used
14. Reserved
15. Reserved
16. Reserved
17. Reserved
18. Reserved
19. Remote release
20. Remote control
21. Ready
22. Ring indicator 1
23. Ring indicator 2
24. Not used
25. Not used

Figure 9-3. *Interface diagram and connections for a typical serial data set. (Courtesy of AT & T)*

FG *Frame Ground* — electrically bonded to the machine or equipment frame.

SD *Send Data* — signals on this circuit are generated by the business-machine equipment and are connected to the transmitting data set for transmission to remote data processing terminal equipment.

RD *Receive Data* — signals on this circuit are generated by the receiving data set in response to data signals received from remote data processing terminal equipment.

RS *Request to Send* — generated by the business machine to condition the local data set to transmit. The data set's carrier signal is transmitted during the ON condition of this circuit, and the ON condition is maintained by the business machine whenever it has information ready for transmission or being transmitted.

CS *Clear to Send* — generated by the transmitting data set to indicate that it is prepared to transmit data. The ON condition is a response to the ON condition of the RS circuit, delayed as may be appropriate by the data set for establishing a communication channel to a remote data processing terminal.

IN *Interlock* — an ON signal indicates to the business machine that the data set has power on, is in the data mode, and is connected to a transmission line. On receipt of Interlock ON the transmitting business machine should turn RS ON indicating to the data set that the business machine wants to transmit.

SG *Signal Ground* — establishes the common ground reference potential for all interface circuits except Frame Ground. Its strapping to Frame Ground is optional as may be required to meet applicable regulations or to minimize the introduction of noise into electronic circuitry.

CO *Carrier On-Off* — used to provide an indication from the data set to the associated business machine that the data carrier is being received. The ON condition indicates reception of the data carrier; the OFF condition provides an indication of the end of present transmission activity or a fault condition.

RI *Ring Indicator* — if the data set is prepared to answer unattended and it receives a call, it will tell the business machine that the phone is ringing by making a closure between the Ring Indicator leads for each ring cycle.

RR *Remote Release* — the business machine indicates that it is prepared to receive a call by making a closure between the RY lead and the RC lead and by keeping a closure between the RR and RC leads. Once the call has been answered, it is not necessary to keep the closure between RY and RC, but it is necessary to keep the closure between RR and RC until it is desired to release the call.

RC *Remote Control* — see discussion for RR.

RY *Ready* — generated by the local data set to indicate that it is ready to operate. See discussion for RR.

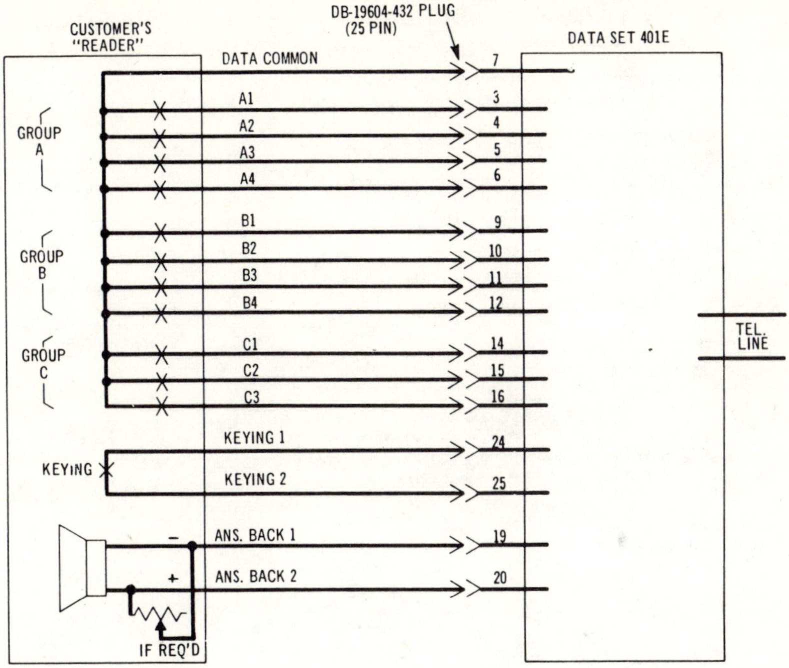


**1.** Reserved
**2.** Reserved
**3.** Data channel A1
**4.** Data channel A2
**5.** Data channel A3
**6.** Data channel A4
**7.** Data common (COM)
**8.** Reserved
**9.** Data channel B1
**10.** Data channel B2
**11.** Data channel B3
**12.** Data channel B4
**13.** Reserved
**14.** Data channel C1
**15.** Data channel C2
**16.** Data channel C3
**17.** Reserved
**18.** Reserved
**19.** Audible answer-back 1 (−)
**20.** Audible answer-back 2 (+)
**21.** Reserved
**22.** Reserved
**23.** Reserved
**24.** Keying — 1
**25.** Keying — 2

Figure 9-4. *Interface diagram and connections for a typical parallel data set. (Courtesy of AT & T)*

Since the data set illustrated is an asynchronous set, it does not contain timing leads in its interface. Synchronous sets, however, use the interface circuits described below.

BT *Business-machine Transmit Timing* — used to provide the transmitting data set with signal element timing information from the business machine. The waveform is nominally ON and OFF for equal periods of time, and a transition from ON to OFF nominally indicates the center of each signal element on the SD circuit.

DT *Data-set Transmit Timing* — same as above, but the data set is the source of the timing information.

**BR** *Business-machine Receive Timing* — same as BT, except that the business machine is supplying information to the data set concerning the timing of the received signal.

**DR** *Data-set Receive Timing* — same as BR, except that the data set is supplying the timing information.

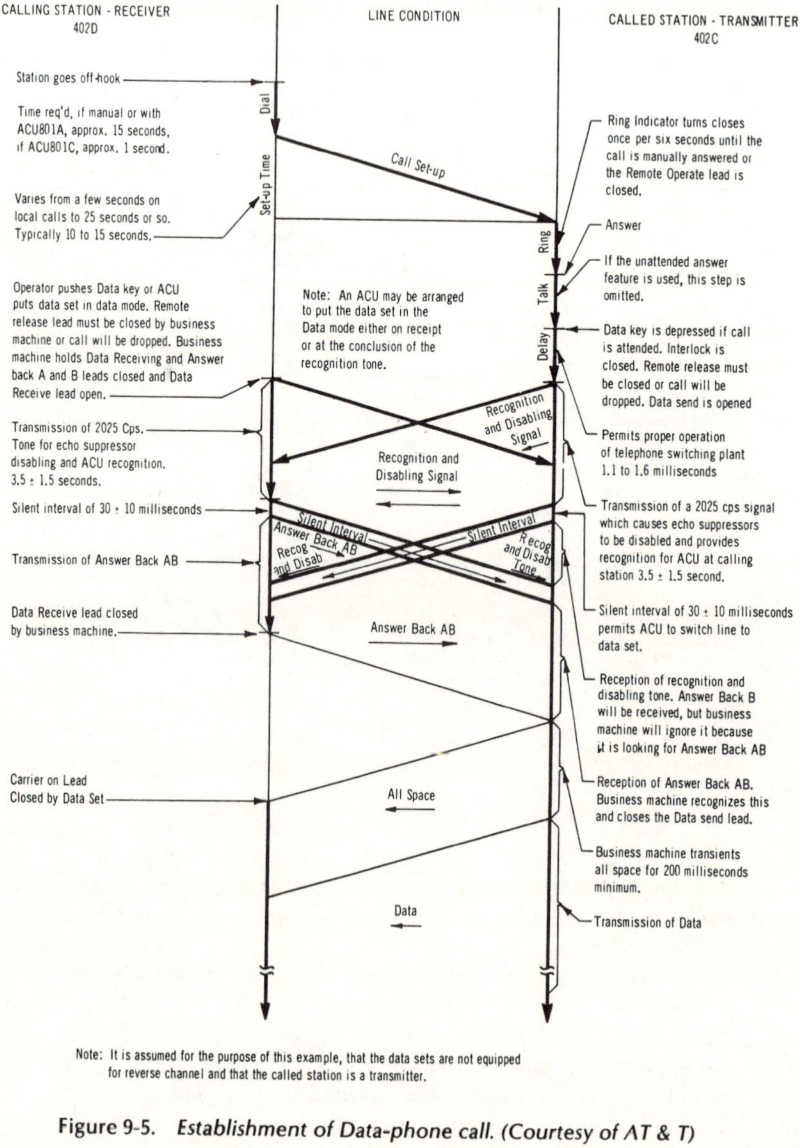


**Figure 9-5.** *Establishment of Data-phone call. (Courtesy of AT & T)*

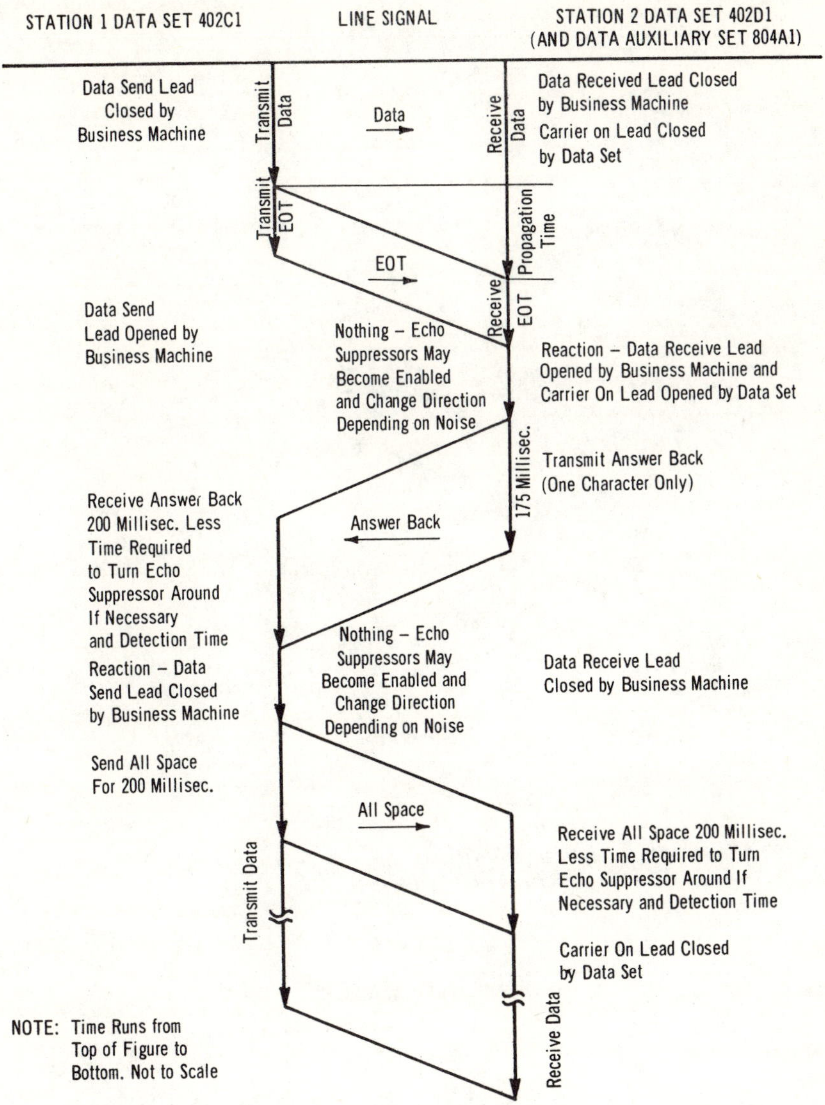


**Figure 9-6.** *Turn-around in Data-phone service station not equipped with reverse channel. (Courtesy of AT & T)*

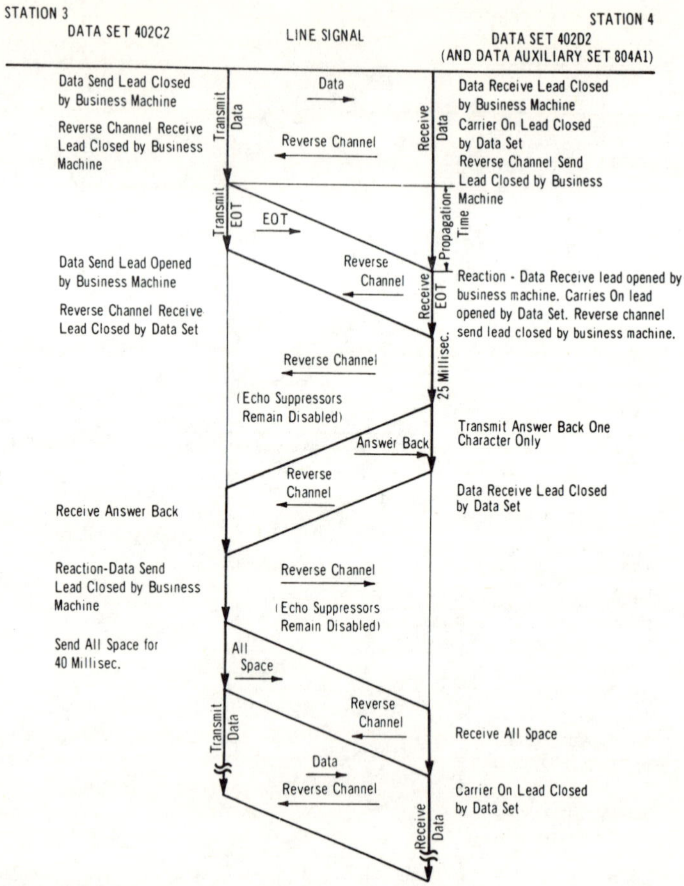


Figure 9-7. *Run-around in Data-phone service stations equipped with reverse channel. (Courtesy of AT & T)*

## DIALING DEVICES

### Rotary

The most familiar device used to place a call through the switched telephone network is the rotary dial found on most telephones. The telephone is connected to the switching equipment when the handset is taken from its cradle (goes "off hook"). As the dial returns to its normal "rest" position after being rotated, it causes d-c pulses to be transmitted to the switching equipment at the rate of about 10 per sec.

### Touch-Tone

A 10-key pad resembling a 10-key adding machine is beginning to replace the familiar rotary dial in some towns. Rather than the d-c pulses of the rotary dial, the keys of the Touch-Tone pad produce multifrequency tones which cause the switching equipment to react. Certain data sets will accept these tones, so it is possible to use Touch-Tone as a data input medium.

### Card Dialer

An automatic dialer which uses a plastic card about the size of a business card is available as optional equipment for telephones and teletypewriters. The card, which is perforated to encode a telephone number is placed in a slot near the dial, and depressed. When a bar is touched by the person dialing, the card travels upward, causing contacts to simulate the pulses of a rotary dial or the multifrequency (MF) tones of a Touch-Tone pad.

### Repertory Dialers

Several manufacturers have built automatic dialers that are capable of storing up to hundreds of telephone numbers. These machines are used in conjunction with ordinary telephone sets and are connected directly to the telephone line. The user of a typical repertory dialer rotates a knob until the name he wishes to call appears in a small window. The associated telephone is then taken off hook, a button on the dialer is depressed, and the telephone number is dialed automatically. Numbers are put into the machine or changed by use of the dial on the associated telephone.

### Automatic Calling Units

It might be desirable to poll many data transmitters or receivers over telephone lines during evening hours. If the data were to be used or generated by a computer, it might be efficient to have the computer do the calling. Automatic calling units (ACU) have been designed to enable computers and other business machines to do this.

Under direction of the associated business machine, the ACU places the telephone line in an off-hook condition and notifies the business machine it is ready to accept the number to be dialed. The number is given to the ACU one digit at a time in binary code. The ACU translates this to the number of DC pulses corresponding to the number being dialed. When the distant station answers, control is given to the business machine for the exchange of data.

## Problems

1. List eight data-set parameters.
2. Compare the output (line signal) of parallel and serial data sets.
3. What is the difference between synchronous and asynchronous data sets?
4. Why can data sets used with leased voice-grade lines be operated at higher speeds than those used with the dialed network?
5. What is a reverse channel when used with a data set?
6. Define interface.
7. What is the function of the interface lead called "Request to Send"?
8. Define "internal timing."
9. List the Bell System data-set series designations and give their characteristics.
10. What is an ACU and what is it used for?
11. What is Touch-Tone? List several applications.

# Glossary

The terms appearing in this glossary are those that are not extensively defined in text. Check the index for entries not listed here but defined in the book.

All terms are listed by key word. "*Bandpass Filter,*" for example, is listed under "*Filter, Bandpass.*" Cross-referencing provides links to synonyms, similarities, and related and contrasting terms.

Some terms have more than one connotation, and an effort has been made to include all accepted definitions found in common usage and in other glossaries. Some definitions have been taken from the Proposed American Standard Vocabulary for Automatic Data Processing. Since these and all other definitions are subject to change for a variety of reasons, those included in this glossary should not be interpreted as final, but as reflecting current usage.

*Accounting, Message:* A system, maintained by a switching device, which records statistical data concerning the messages it handles; e.g., number of messages to and from each station, message lengths, queue lengths, etc.

*Accuracy:* The degree of freedom from error; e.g., a six-place table containing an error is more precise but less accurate than a four-place table that is error-free.

*Acknowledgment:* A character or group of characters generated at a receiving device to indicate to the sending device that information has been received either correctly or incorrectly.

*Address, Group:* A coded representation of several stations. A group address is used to activate a certain group of stations simultaneously. Synonymous with *Address, Multiple.* Contrasts with *Broadcast.*

*Address, Multiple:* See *Address, Group.*

*Address, Single:* A code that specifies a single station that is to receive traffic.

*Alphanumeric:* 1) A contraction of alphabetic-numeric. 2) Pertaining to a character set that contains both letters and numerals, and usually special characters.

*Amplification:* 1) The strengthening of a weak signal. Contrasts with attenuation. 2) The ratio between the output signal power and the input signal power of a device. 3) Gain.

*Amplitude:* The strength of a signal. Amplitude and time are the two parameters of any signal.

*Analog:* Pertaining to data in the form of continuously variable physical quantities. Contrasts with *Digital.*

*Answerback:* 1) A signal or sequence of signals from a receiving business machine indicating that it is ready to accept or acknowledge receipt of data. The signal may be received audibly through a loudspeaker at the transmitting end for manual operation, or electronically for automatic operation. 2) The act of transmitting a signal as described in 1).

*Area Numbering Plan (NPA):* The unique scheme of uniform numbers (area codes) assigned to geographical locations served by the national switched telephone network.

*Audio:* Frequencies that can be heard by the human ear, usually between 20 and 20,000 Hz.

*Band:* 1) A range of frequencies between two defined limits. 2) A group of recording tracks on a magnetic drum. 3) One of seven geographical WATS calling areas.

*Bandpass Filter:* See *Filter, Bandpass.*

*Bandwidth, Nominal:* The maximum range of frequencies, including guard bands, assigned to a channel. Only the range between the guard bands is usable for data transmission under most circumstances.

*Baseband:* 1) The frequency band occupied by the aggregate of the transmitted signals before they have modulated a carrier. 2) A signal in the frequency band of 1).

*Bias:* The degree of displacement of the space-to-mark (or vice versa) transition in a received binary signal. Measured as a percentage of the original (or optimum) transition intervals. Also see *Margin.*

*Binary-Coded Decimal:* See *Decimal, Binary-Coded.*

*Biquinary:* Pertaining to the number representation system in which each decimal digit N is represented by the digit pair, AB, where N = 5A + B, A = 0 or 1, and B = 0, 1, 2, 3, or 4.

*Bi-Stable:* The capability of assuming either of two stable states, thus of storing one bit of information.

*Bit, Start:* A bit preceding a character in certain serial transmissions; used to permit synchronization of the transmitter and receiver.

*Bit, Stop:* A bit following a character in certain serial transmissions; used to reset the receiver in preparation for receipt of the next character.

*Blank:* A condition usually represented by no perforation in a given character location on a punched tape or card. Sometimes used as a control character.

*Block:* 1) A set of things, such as words, characters, or digits, handled as a unit. 2) A circuit assemblage that functions as a unit.

*Bound:* To be limited in speed by a particular component of a system. A system that is line-bound, for example, is limited in performance by the speed at which information can be transmitted over the communications line. A printer-bound system has a printer as its slowest component under the specified circumstances.

*bps:* bits per second.

*Broadband:* 1) Same as wideband. 2) Western Union's Broadband Exchange Service.

*Broadcast:* To transmit to all stations on a circuit, as opposed to one or a particular group of stations. Also, see *Address, Group* and *Address, Single.*

*Buffer:* 1) A storage device used to compensate for a difference in rate of flow of data, or time of occurrence of events, when data is transmitted from one device to another. 2) A system component which permits a change of speed, voltage, code structure, or interface characteristics; frequently presumed to include storage capability.

*Capacity, Circuit:* 1) The number of communications channels that can be derived from a given circuit at the same time. 2) The information capacity, measured in bits per second, of a circuit.

*Card, Edge-Punched:* A card of fixed size (may be fanfolded) into which information may be recorded or stored by punching holes along one edge in a pattern similar to that used by punched tape.

*Card, 80-Column:* A punched card with 80 vertical columns representing 80 characters. Each column is divided into two sections, one with character positions labeled 0 through 9, and the other labeled 11 and 12. The 11 and 12 positions are also referred to as the X and Y zone punches, respectively.

*Card, Punched:* 1) A heavy stiff paper of constant size and shape, suitable for punching in a pattern that has meaning, and for being handled mechanically. 2) Also see *Card, Edge-Punched.*

*CCITT:* International Telegraph and Telephone Consultative Committee, an agency of the United Nations International Telecommunications Union. Its purpose is to promote compatibility between communications practices and performance standards of various nations. CCIR performs a similar function.

*CDC:* Call-Directing Code. See *Code, Call-Directing.*

*Center, Rate:* A defined geographical location used for measuring mileage for long-distance and private-line rate purposes. Rate centers are listed in FCC Tariff No. 245.

*Center, Torn-Tape Switching:* A message-switching activity where operators tear off the incoming punched and printed paper tape and transfer it manually to the proper outgoing circuit.

*Central Processing Unit:* See *Unit, Central Processing.*

*Channel, Analog:* Usually refers to a channel that will pass alternating current, but not direct current. A switched voice channel is an analog channel, while most teletypewriter circuits are digital (d-c) channels. If an analog channel is said to carry digital data, it is actually carrying analog representations of the digital data in the form of various frequencies.

*Channel, Class of:* Refers to the media through which channels pass: wire, cable, carrier, coaxial cable, radio, and microwave.

*Channel, Digital:* A channel capable of carrying direct current, as opposed to analog channels, which do not.

*Channel, Grade of:* Refers to the relative bandwidth of a channel: narrowband, voiceband, wideband.

*Channel, Local:* The wire link, in a telephone system, extending from the central office to the subscriber's premises.

*Channel, Subvoice:* A channel with less bandwidth than a normal voice channel. Also called a narrowband channel.

*Channel, Tape:* See *Level, Tape.*

*Channel, Telegraph-Grade:* 1) A communications path suitable for transmission of signals at speeds up to approximately 180 bauds, but not suitable for voice due to restricted bandwidth. 2) A narrowband channel.

*Channel, Telephone:* See *Channel, Voice-Grade.*

*Channel, Type of:* A channel type refers to the way in which it is used: simplex, half-duplex, or full duplex.

*Channel, Voice-Grade:* 1) A common telephone connection between points. May be a private (leased) line or made through the switched network. Same as telephone channel. 2) A channel suitable for transmission of speech, digital or analog data, or facsimile, generally with a frequency range between 300 and 3300 Hz.

*Channelize:* To subdivide a channel into multiple channels of lesser bandwidth.

*Character, Check:* A parity character added to a group of characters to assist in error detection-correction.

*Character, Control:* 1) A character whose occurrence in a particular context initiates, modifies, or stops a control operation or function. 2) A character used to initiate functions such as line feed, carriage return, etc. 3) Synonymous with function code.

*Character, Parity:* A column of parity bits representing the modulus-2 sum (binary) of each horizontal row of bits in a medium.

*Character, Special:* In a character set, a character that is neither a numeral nor a letter.

*Characteristics, Signal:* The basic characteristics of a signal wave form are its amplitude, frequency, and phase.

*Check, Cyclic:* A method of error detection which checks every nth bit, n + 1 bit, n + 2 bit, and so forth. It is more powerful and efficient than horizontal checks, vertical checks, or combinations of both.

*Check Digit:* See *Digit, Check.*

*Circuit, Dialed:* A connection established through the switched telephone or message network by dialing or by keying a tone transmitter. Such circuits can be established manually or automatically by a business machine.

*Circuit, Equalized:* A circuit in which the frequency and/or phase distortion has been reduced to prescribed limits by the introduction of networks to compensate for the difference in attenuation and/or time delay at the various frequencies in the transmission band.

*Circuit, Four-Wire:* A system in which the transmitting and receiving paths are separate channels.

*Circuit, Leased:* A service offering which provides a customer with permanent (rather than dialed) connections to all points on the circuit for the duration of a contract.

*Circuit, Multipoint:* A system consisting of a circuit connecting three or more terminals, any or all of which may simultaneously receive information flowing in the common circuit.

*Circuit, Phantom:* A third circuit derived from two physical circuits by means of repeating coils installed at the terminals of the physical (side) circuits.

*Circuit, Physical:* A metallic two-wire channel.

*Circuit, Two-Wire:* A system in which all transmitting and receiving is performed over one pair of wires (or equivalent).

*CO:* Central Office. See *Office, Central.*

*Code, Binary:* Any code using only two distinguishable code elements or states, representing the binary digits 1 and 0.

*Code, Call-Directing (CDC):* A two- or three-character code used in teletypewriter systems to activate a particular machine or group of machines.

*Code, Column-Binary:* A code used with punched cards, in which successive bits are represented by the presence or absence of punches, in binary notation, in successive columns (as opposed to rows).

*Code, Field Data:* The U.S. military code used as a compromise between conflicting manufacturer's codes.

*Code, Five-Level:* 1) A term frequently used as a synonym for baudot code. 2) Any code using five elements, or bits, to designate one character. Such a code has 32 possible discrete combinations of the five code elements.

*Code, Forward-Acting:* An error-control code that contains redundancy added in such a way that one or more characters received in error can be reconstructed at the receiving end without additional information from the source.

*Code, Function:* See *Character, Control.*

*Code, Insensitive:* See *Transparent.*

*Code, Machine:* 1) A set of binary digits used by a business machine. 2) An operation code.

*Code, Morse:* A system of dot-dash signals invented by Alfred Vail. Not used for data communications.

*Code, n-ary:* A code in which each code element may be any one of n distinct values or kinds. For example, the code elements of a *bi*nary code have *two* different values; quaternary code elements have four possible values, etc.

*Code, Operation:* One of a set of instructions that causes a machine to perform an operation other than its main function. For example, the main function of a receiving teletypewriter is to print characters, but it can also perform operations such as line feed, carriage return, bell, etc.

*Code, Transmitter-Start (TSC):* The unique code sent to a particular teletypewriter on a circuit to cause it to transmit its tape.

*Column:* A vertical arrangement of characters or other expressions. Contrasts with *Row.*

*Communications:* The means of conveying information of any kind from one person or place to another, except by direct unassisted conversation or correspondence.

*Computer:* Any device capable of accepting information, applying prescribed processes to the information, and supplying the results of these processes.

*Conditioning, Channel:* The electrical balancing of a channel to reduce attenuation distortion, delay distortion, and effects of noise.

*Control, Error:* 1) An arrangement that will detect the presence of errors. In some systems, refinements are added that will correct the detected errors, either by operations on received data (see *Code, Forward-Acting)* or by retransmission from the source. 2) A periodic or continuous check of such channel characteristics as delay distortion, frequency response, and noise in an effort to reduce or eliminate the causes of errors.

*Control, Line:* 1) A system or method of presenting incoming information to the central processor and outgoing information to the terminal lines in a manner that (ideally) optimizes the utilization of the hardware and software involved (for computer systems). 2) Sets of procedures that designate how line usage will be controlled.

*Core, Magnetic:* 1) Commonly, the memory unit contained as an integral part of the central processing unit in a computer. 2) A configuration of magnetic material that is placed in a spatial relationship to current-carrying conductors, and whose magnetic properties are essential to its use. It may be used to retain a magnetic polarization for the purpose of storing data, or for its nonlinear properties as in a logic element.

*cps:* 1) Characters per second. 2) Cycles per second. Now referred to as hertz (Hz).

*CPU:* Central Processing Unit of a computer. See *Unit, Central Processing.*

*Crossbar:* An automatic telephone switching system using switches mounted on movable bars. Information dialed into the system is stored in common equipment that tests and selects the switching paths and controls the operation of the switching mechanism.

*Cycle:* 1) In alternating current, change of value through maximum positive and negative values and back to the original value. The number of cycles per second is the frequency, which is stated in units of hertz. 2) The interval of one completion in a recurrent process. 3) To repeat a set of operations indefinitely or until a stated condition is met.

*Decimal:* 1) The numeration system with a base of ten. 2) A characteristic or property involving a selection, choice, or condition in which there are ten possibilities.

*Decimal, Binary-Coded:* Pertaining to a decimal notation in which the individual decimal digits are each represented by a group of four binary digits. For example, the decimal number 23 would be represented as 10111 in pure binary notation, but as 0010 0011 in binary-coded decimal notation. Abbreviated as BCD.

*Deck:* A collection of cards; commonly a complete set of cards which have been punched for a definite service or purpose.

*Decode:* To determine the meaning of individual characters or groups of characters in a message through the reversal of some previous coding.

*Dedicated:* Synonymous with leased or private. Usually refers to communications lines or equipment.

*Delay:* The difference in time between a cause and its effect; e.g., the transmission and receipt of a pulse.

*Delay, Absolute:* The real-time interval from the transmission to reception of a signal over a circuit. Also called transmission time or circuit delay.

*Delay, Circuit:* See *Delay, Absolute.*

*Density:* The number of units of useful information contained within a linear dimension, usually expressed in units per inch.

*Dial-Up:* 1) The use of a dial or pushbutton telephone to initiate a station-to-station telephone connection. 2) The network that provides the interconnection for calls as in 1).

*Digit, Check:* A redundant digit (or digits) carried within a unit item of information (character, word, block, etc.) which provides information about the other digits in the unit in such a manner that if an error occurs the check fails. Also see *Parity.*

*Digital:* 1) Data in discrete quantities. Contrasts with *Analog.* 2) Pertaining to data in the form of digits.

*Disk, Magnetic:* A storage device on which information is recorded on the magnetizable surface of a rotating disk. A magnetic disk storage system is an array of such devices with associated reading and writing heads mounted on movable arms.

*Display, Visual:* A provision, such as a cathode-ray tube, for the visual representation of data.

*Distortion, Peak:* Worst time displacement of the transmission wave components. Also called jitter.

*Document, Source:* A form, voucher, or written evidence of a transaction from which basic data is extracted.

*Drop:* 1) An intermediate terminal point between the ends of a multipoint channel. 2) The telephone wire between a utility pole and a customer's building.

*Edit:* To rearrange, delete, or add information. For example, blanks, spaces, or invariant symbols may be removed from data to be transmitted to allow more efficient use of available circuit time, then replaced after the data has been received.

*EDP:* Electronic Data Processing.

*Electronic:* Pertaining to that branch of science dealing with the motion, emission, and behavior of currents of free electrons, especially in vacuum, gas, or phototubes and special conductors or semi-conductors. This is contrasted with *electric,* which pertains to the flow of large currents in metal conductors.

*Element, Signal:* 1) That part of a signal that occupies the shortest interval of the signaling code. 2) The unit duration in building up signal combinations. 3) A pulse, or signal. An absence or presence of voltage or current in a communications medium.

*Encode:* Conversion of a character into its equivalent combination of bits.

*EOA:* End Of Address.

*EOM:* End Of Message.

*Equipment:* A general term meaning machinery, either collectively or in particular.

*Equipment, Station:* Usually telephone or teletypewriter apparatus, but could include business-machine units.

*Erase:* 1) To replace all the binary digits in a magnetic storage device by binary zeros. 2) To replace all the binary digits in a paper tape by punched holes; more correctly called rubout or letter out.

*Error:* Any discrepancy between a computed, observed, recorded, or measured quantity and the true, specified, or theoritcally correct value or condition. Contrasts with *Mistake.*

*Error, Conscious:* An error that was instantly recognized as such by an operator, but that his reflex actions were unable to prevent.

*Error, Line:* A deviation from correctness in data caused by conditions in a transmission medium.

*Exchange:* 1) A defined area, served by one or more central offices of a communications common carrier, within which the carrier furnishes service at the exchange rate and under the regulations applicable in that area prescribed in the carrier's filed tariffs. 2) See *Office, Central.*

*Facilities, Exchange:* Channels and equipment used by the communications common carriers to provide service to the general public. In a broader sense, any such channels and equipment. Contrasts with *Facilities, Private.*

*Facilities, Leased:* See *Facilities, Private.*

*Facilities, Private:* Channels and equipment used by the communications common carriers to provide service dedicated to the use of one subscriber exclusively. Contrasts with *Facilities, Exchange,* and synonymous with *Facilities, Leased.*

*Fading:* The fluctuating in intensity of any or all components of a received radio signal due to changes in the characteristics of the propagation path.

*FDX:* Full duplex.

*FEX:* Foreign Exchange Service (often FX).

*Feed:* 1) To supply the material to be operated upon to a machine. 2) A device capable of feeding, e.g., a tape feed or a card fee.

*Feed, Friction:* A teletypewriter term meaning a device similar to a typewriter platen used to advance the paper. Cannot be used with printed forms due to lack of positive paper positioning. See *Feed, Sprocket.*

*Feed, Sprocket:* 1) A machine having a platen equipped with sprockets that engage holes punched along both edges of continuous forms for accurate paper positioning. 2) The forms used in the above machine. See *Feed, Friction.*

*Feedhole:* One of a series of non-intelligence holes in punched paper tape which engage with a driving sprocket to permit transportation of the tape over sensing or punching devices.

*Filter, Bandpass:* A wave filter that blocks all frequencies outside a designated bandwidth, but does not inhibit those frequencies within the bandwidth(s). Neither cutoff frequency is zero or infinity.

*Flip-Flop:* A bistable device capable of assuming either of two stable states and of storing one bit (either a 1 or 0) of information. Often used to establish parity.

*Frame, Main:* The central processor of a computer system, containing the main storage, the arithmetic unit, and special register groups.

*Frequency, Carrier:* The number of hertz of the carrier signal.

*Frequency, Voice:* 1) That portion of the audio frequency range which is used to transmit speech and other information; usually between the limits of 300 and 3300 Hz. 2) The range of the human voice, which is approximately between 80 and 1180 Hz.

*Function, Control:* 1) A nonprinting bit configuration that causes a machine action such as line feed, carriage return, etc., but has no affect on data. 2) The machine operation caused by 1).

*Grade, Channel:* The grades of channels are wideband, voiceband, and narrowband. See *Type, Channel.*

*Graphic:* A written or printed letter or symbol.

*Guardband:* An unused frequency band between two channels to give a margin of safety against mutual interference.

*Hardware:* A colloquialism commonly applied to the mechanical, electrical, and electronic features of a system. It is most frequently used to differentiate between the physical features of the system and those which the user introduces through coding or programming. Contrasts with *Software.*

*Harmonic:* A sinusoidal component of a periodic wave, having a frequency that is an integral multiple of the fundamental frequency.

*Hartley:* A choice between ten equiprobable events.

*Head, Read-Write:* A small electromagnet used for reading, recording, or erasing polarized spots, which represent information, on magnetic tape, disk, or drum.

*Hexadecimal:* 1) A characteristic property involving a selection, choice, or condition in which there are 16 possibilities. 2) The numeration system with a base of 16.

*Hit, Line:* An electrical interference of very short duration causing the introduction of unwanted signals on a circuit. Also called a spike.

*Image, Card:* A representation in storage of the holes punched in a card, in such a manner that each card hole is represented by one binary digit and the unpunched spaces are each represented by the other binary digit. Clarified by *Code, Column Binary.*

*Impedance:* The total opposition offered by a component or circuit to the flow of an alternating or varying current; a combination of resistance and reactance.

*Initialization:* The process and/or time required to establish initial contact between transmitter and receiver on a circuit. Contrasts with *Resynchronization.*

*Input:* 1) Data to be processed. 2) The device or devices used to bring data into another device. 3) The process of transferring data from an internal or local device. Can be automatic or manual. 4) Paper tape, punched cards, or other media containing data to be transferred to a distant location.

*Inquiry:* A technique whereby the interrogation of the contents of a computer's storage may be initiated at a local or remote point by use of a keyboard, touchtone pad, or other device.

*Intelligence:* 1) Knowledge of information. 2) Capacity for understanding; aptitude in grasping truths, facts, meaning; the faculty for understanding.

*Interference, Adjacent Channel:* The presence of undesirable energy in a channel caused by one or both sidebands of modulated carrier channels in close frequency proximity extending from one into another through electrostatic or electromagnetic coupling.

*Interleave:* To send blocks of data alternately to two or more stations on a multipoint circuit.

*Jitter:* A shift in the time or phase position of individual pulses, causing difficulty in synchronization and/or detection. Also called peak distortion.

*kc:* kilocycle; 1000 cycles per second. Now called kilohertz (kHz).

*Key:* 1) One of a set of levers or parts pressed to operate a teletypewriter, console, etc. 2) To depress a lever [ as in 1) ], thus causing an associated function to occur.

*KSR:* Keyboard Send and Receive. A teletypewriter equipped with keyboard and printer, but with no tape-handling capabilities.

*Language:* A set of representations, conventions, and rules used to convey information between people, or between people and machines.

*LD:* Long Distance.

*Level, Decision:* A predetermined signal amplitude that serves as a standard for determining the value of a signal pulse. If the signal amplitude is above the decision level at the time of sampling, a binary 1 is indicated. A binary 0 is indicated if the signal amplitude is below the decision level at sampling time. Ideally, a signal pulse is measured at its time center. Also called decision threshold and slicing level.

*Level, Priority:* The degree of precedence or urgency assigned to a message or operation.

*Level, Signal:* 1) An expression of relative signal strength at various points in a communication circuit. 2) In telephone, the signal strength relative to an established reference signal known as zero reference level, usually understood to be at the point where the circuit being considered becomes a toll circuit. See *Power, Signal* for contrast.

*Level, Tape:* A row or channel of punched or unpunched spaces parallel to the edge of a paper tape, usually ten to the inch. Also see *Tape, Punched Paper.*

*Line:* 1) A pair of wires. 2) A telephone circuit. 3) A row of characters on a printed page.

*Link, Communication:* 1) The means of connecting one location to another for the purpose of transmitting and receiving information. 2) A channel or circuit intended to connect other channels or circuits.

*Loop, Local:* That part of a communication circuit between the customer's premise or to the central office (either two-wire or four-wire.)

*LPM:* Lines Per Minute.

*LRC:* Longitudinal Redundancy Check.

*Machine, Business:* In communications, a device capable of creating or absorbing information in machine language. This action may be in conjunction with some other function, such as calculating, measuring, etc.

*Machine-Sensible:* In machine language, or capable of being interpreted by a machine.

*Magnitude, Orders of:* The Institute of Electrical and Electronics Engineers (IEEE) recently adopted some new standards of units and symbols, some of which are reflected in the first column of the table on top of the following page.

*Malfunction:* An equipment failure. Contrasts with *Mistake* and *Error.*

*Margin:* That fraction of a perfect signal element through which the time of selection may be varied in one direction from the normal time of selection, without causing errors while signals are being received. There are two distinct margins, determined by varying the time of selection in either direction from normal. See *Element, Signal* and *Bias.*

*Mc:* Megacycle; 1,000,000 cycles per second. Now called megahertz (MHz).

*Memory:* See *Storage.*

*Message, Multiple-Address:* A message that is destined for two or more addressees, each of whom is informed of all the addressees (by the information in the heading).

*Message, Service:* A transmission between communications peronnel pertaining to any phase of traffic

| | | | | |
|---|---|---|---|---|
| T | tera | 1,000,000,000,000 | $10^{12}$ | million million |
| G | giga | 1,000,000,000 | $10^{9}$ | thousand million |
| M | mega | 1,000,000 | $10^{6}$ | million |
| k | kilo | 1000 | $10^{3}$ | thousand |
| h | hecto | 100 | $10^{2}$ | hundred |
| da | deka | 10 | 10 | ten |
| | | 0 | | |
| d | deci | 0.1 | $10^{-1}$ | tenth |
| c | centi | 0.01 | $10^{-2}$ | hundredth |
| m | milli | 0.001 | $10^{-3}$ | thousandth |
| $\mu$ | micro | 0.000,0001 | $10^{-6}$ | millionth |
| n | nano | 0.000,000,001 | $10^{-9}$ | thousand millionth |
| p | pico | 0.000,000,000,001 | $10^{-12}$ | million millionth |
| f | femto | 0.000,000,000,000,001 | $10^{-15}$ | thousand million millionth |
| a | atto | 0.000,000,000,000,000,001 | $10^{-18}$ | million million millionth |

handling, communication facilities, or circuit condition.

*Microsecond:* One millionth of a second. One microsecond is to 1 second as 1 second is to 12 days.

*Millisecond:* One thousandth of a second. One millisecond is to 1 second as 1 second is to 17 minutes.

*Mistake:* A human blunder or failing. Contrasts with *Error* and *Malfunction.*

*MIS:* Management Information System. See *System, Management Information.*

*Mode:* A condition or method of operation.

*Mode, Burst:* The movement of a continuous bit stream between devices until an interruption or completion of the stream occurs. Compare with *Mode, Byte.*

*Mode, Byte:* The movement of one byte at a time between devices, separated by an interrupt and release of channel control. Used in multiplexing, the byte mode permits the handling of data from several low-speed devices simultaneously.

*Mode, Conversation:* A condition of real-time communication between one or more remote terminals and a time-sharing computer, in which each entry from a terminal elicits an immediate response from the computer. The remote terminal thus can control, interrogate, or modify a task within the computer. Contrasts with *Mode, Reactive.*

*Mode, Reactive:* A condition of communication between one or more remote terminals and a computer, in which each entry (usually batch) causes certain actions to be performed by the computer, but not necessarily including an immediate reply. Contrasts with *Mode, Conversation.*

*Modem:* A contraction of *modu*lator/*dem*odulator. A data set that both transmits and receives.

*Modulation, Angle:* A modulation method in which the angle of the sine wave is the carrier's characteristic that is varied in accordance with the impressed intelligence signal. Two forms of angle modulation are phase modulation (PM) and frequency modulation (FM).

*Modulation, Frequency-Shift (FSM):* A form of frequency modulation in which the carrier frequency is shifted between a mark frequency and a space frequency in response to the impressed intelligence signal (frequency-shift keying).

*Modulation, Pulse Code (PCM):* Modulation in which a coded pulse is used to represent quantized values of instantaneous samples of the signal wave. The characteristics of pulses within a train may be modified in one of several ways to convey information, including pulse amplitude (PAM), pulse position (PPM), and pulse duration (PDM).

*Module:* An interchangeable and/or incremental item containing components.

*Monitor:* 1) To check operation and performance of a system or circuit by examining parts of transmissions. 2) A station or equipment arranged to supervise system operation.

*Nanosecond:* One billionth of a second. One nanosecond is to 1 second as 1 second is to 32 years.

*Network, Exchange:* The public switched telephone system. Often referred to as the message network, the toll system, the DDD network, or the telephone network. Contrasts with *Network, Leased.*

*Network, Leased:* A multipoint telephone system utilizing lines and equipment leased from a telephone company or other common carrier, and reserved for the exclusive use of one customer. Contrasts with *Network, Exchange.*

*Network, Message:* Same as *Network, Exchange.* Sometimes used to refer to the dialed teletypewriter network as opposed to the dialed telephone system.

*Network, Private-Line:* Same as *Network, Leased.*

*Network, Switched Message:* Same as *Network, Exchange.* Sometimes used to refer to the dialed teletypewriter network as opposed to the dialed telephone system.

*Network, Switched Voice:* Same as *Network, Exchange.*

*Network, Telephone:* A telephone system composed of multiple, as opposed to 2-point, lines and sets. Usually implies switching capabilities. Can mean public or private (leased) systems. *A* telephone network is as described above; *the* telephone network implies the nationwide public switched telephone system.

*New Sync:* See *Sync, New.*

*Noise, Ambient:* Extraneous frequencies in the audio range. These can become troublesome in systems using acoustical couplings.

*Noise, Line:* Unwanted signals originating in a channel.

*Nominal Bandwidth:* See *Bandwidth, Nominal.*

*Notation:* The act, process, or method of representing facts or quantities by a system or set of marks, signs, figures, or characters.

*Numeral:* A digit, character, or symbol used to represent a quantity.

*Office, Central:* A telephone switching office where lines serving customers in a limited geographical area are terminated, interconnected, or connected to toll facilities. Usually limited to serving about 10,000 telephones.

*Online:* Equipment or operations that are in direct contact with other points on a circuit. An online teleprocessing system eliminates the need for human intervention between information input at the source and ultimate processing by a computer. Contrasts with *Real-Time.*

*Open:* A condition in which conductors are separated so that current cannot pass.

*Order:* Pertaining to the weight or significance assigned to digit position in a number.

*Order, High:* The digit farthest to the left within a number, e.g., in the number 2345, the high-order digit is 2.

*Order, Low:* The digit farthest to the right within a number, e.g., in the number 2345, the low-order digit is 5.

*Pack:* To combine two or more units of information into a single physical unit to conserve storage or transmission time.

*Pair:* Two like conductors employed to form an electric circuit.

*Parallel:* The simultaneous transmission or processing of the individual parts of a whole, such as the bits of a character or the characters of a word. When characters are sent sequentially, but the bits within the characters are dealt with simultaneously (not one after another), the transmission is serial by character, parallel by bit.

*Parameter:* A definable characteristic of an item, device, or system.

*Parity:* Equivalence of value in the check digit of the transmitted and the received data.

*Parity, Cyclic:* See *Check, Cyclic.*

*Path:* A circuit or channel.

*PCM:* 1) Pulse Code Modulation. 2) Punched-Card Machine; the machines and equipment using punched cards.

*Phantom Circuit:* See *Circuit, Phantom.*

*Phase:* The position of a point on the waveform of an alternating or other periodic quantity with respect to the start of the cycle, usually expressed in degrees. Also see *Shift, Phase.*

*Plant:* Physical equipment and line facilities installed and operated by the telephone or telegraph companies to provide communications by electrical means.

*Polar:* A situation in which a binary 1 is represented by current flow in one direction and a binary 0 by current flow in the opposite direction.

*Power, Signal:* An expression of absolute signal strength at a specific point in a circuit. See *Level, Signal* for contrast.

*Printer, Line:* A printer in which all characters across an entire line of type are printed in one print cycle. Contrasts with *Printer, Page.*

*Printer, Page:* A printer that prints characters one at a time in page format. Contrasts with *Printer, Line.*

*Printer, Serial:* Same as *Printer, Page.*

*Priority:* Degree of precedence or urgency. Message switching systems usually have the ability to recognize and handle several grades of priority.

*Processing, Centralized Data:* Data processing performed at a single, central location on data obtained from several geographical locations.

*Program:* 1) The complete sequence of machine instructions and routines necessary to solve a problem. 2) To plan the procedures for solving a problem. 3) Software.

*Propagation:* The travel of waves through or along a medium.

*Protection, Message:* Assurance of message delivery. Accomplished by maintenance of message log, check for missing serial numbers, and analyzation of message addresses for accuracy and validity.

*Pulse:* A relatively sharp, short variation in the intensity of an electrical quantity. A pulse is characterized by its amplitude, duration, and rise and decay.

*Quality:* Fidelity of reproduction of an original phenomenon.

*Queue:* 1) A waiting line or area. 2) The contents of a waiting line or area.

*Queuing, Message:* The holding of messages in sequence until the destination terminal can be reached.

*Range:* 1) All the values a function may have. 2) The difference between the highest and lowest values a function may have.

*Rate, Baud:* The number of signal elements that pass through a channel or device in a second. See *Element, Signal.*

*Rate, Sampling:* The frequency at which measurements of physical quantities are made.

*Rate, Signaling:* The number of code elements transmitted per second. Synonymous with baud.

*Read:* To sense the information on a recording medium.

*Reader, Paper-Tape:* A device capable of sensing information punched in a paper tape in the form of a pattern of holes.

*Real-Time:* Performance of a computation or processing during the time of the actual related function in order that the results are available for guiding the actual process of the function. Contrasts with batch processing. See *System, Real-Time.*

*Regeneration:* A method of communication in which a received signal is passed directly to an associated transmitter and is sent to the next station on the circuit. A tap is provided between the receiver and the transmitter to provide an associated business machine with the received signal and to accept output signals from the machine. Effectively, every link in such a system is a point-to-point channel between business machines, as contrasted with a multipoint system.

*Reliability:* The probability that a device or medium will function without failure over a specified time period or amount of usage.

*Reliability, Circuit:* The percentage of time the circuit meets arbitrary standards.

*Repeater:* A transmission system component that reconstitutes communication signals into standard voltages, currents, and/or timing.

*Response, Frequency:* The degree to which different frequencies are attenuated during transmission over a given facility; the variation of the circuit net loss with frequency.

*Resynchronization:* On a multipoint circuit, the reestablishment of synchronization with a receiver (or transmitter) with which contact had recently been made. Contrasts with *Initialization.*

*Retrieval, Information:* The methods and procedures for recovering specific information from stored data. Also see *Retrieval, Message.*

*Retrieval, Message:* The recall of messages from a log or queue, some time after the messages have been handled by a message-switching system.

*Return, Carriage:* In a character-by-character printing mechanism, the operation that causes the next character to be printed at the left margin.

*Route, Alternate:* A secondary communication path used to reach a destination if the primary path is unavailable.

*Row:* 1) A horizontal arrangement of characters, bits, or other expressions. Contrasts with *Column.* 2) If used in conjunction with paper tape, often equated with level or channel.

*Row, Card:* One of the horizontal lines of punching positions on a punched card.

*Scan:* Examine sequentially part by part.

*Sense:* An attempt to detect holes or marks in or on storage media.

*Set, Character:* The numbers, letters, and symbols associated with a given device or coding system.

*Shannon:* A choice between two equiprobable events.

*Shift, Phase:* A change in time relationship of one part of a signal waveform with another, with no change in the basic form of the signal. The degree of change varies with frequency as a signal passes through a channel.

*Skew:* A degree of nonsynchronism of supposedly parallel elements.

*Software:* All programs and routines used to extend the capabilities of computers, such as compilers, assemblers, narrators, routines, and subroutines. Contrasts with *Hardware.*

*SOM:* Start Of Message.

*Speed, Line:* 1) The maximum rate at which signals may be transmitted over a given channel. 2) The normal operating speed of a communications system. Refers to the automatic machine speed, as opposed to the slower speed of any associated manual operations.

*Speed, Transmission:* 1) The rate at which data is transferred. Usually measured in bits per second, characters per second, or words per minute. 2) The absolute time a signal requires to move between source and sink. This approaches the speed of light (about 186,000 miles per second) and is affected by distance, transmission medium, number of system components, etc. Same as *Time, Propagation.*

*Spike:* See *Hit, Line.*

*Start-Stop:* A transmission method employing coded characters made up of a start bit, several (5 to 8) information bits, and a stop bit. The receiving machine stops after each character and starts again upon receipt of the first bit from the next character (the start bit). Synonymous with asynchronous.

*Station:* 1) An input or output point on a communication system; a terminal; a machine. 2) A telephone set. 3) A magnetic tape unit.

*Storage:* A device to which data can be transferred and from which it can be obtained at a later time.

*Stream, Bit:* Referring to a binary signal without regard to groupings by character, word, or other unit. A line signal.

*Subscriber:* A customer of a common carrier.

*Subset:* 1) A data set. 2) Any collection of objects or items contained within a larger set. 3) Contraction of subscriber's set, which is any device (but usually a telephone) installed on a subscriber's premises.

*Supergroup:* A group of 60 voice-grade channels with a bandwidth of 240 Hz.

*Switch:* 1) To establish a temporary interconnection between two or more stations over communications paths. 2) A short term for a line or message switcher.

*Switching, Torn-Tape:* A method of message relay in which operators tear off incoming printed and punched paper tape and transfer it manually to the proper outgoing circuit.

*Symbol:* Synonymous with baud. See *Element, Signal.*

*Sync, New:* A feature of some data sets that allows for a rapid resynchronization during transition from one transmitter to another on a multipoint private-line data network.

*System, Carrier:* A means of obtaining a number of channels over a single circuit, or path, by modulating each channel upon a different carrier frequency and demodulating at the receiving point to restore the signals to their original form.

*System, Management Information (MIS):* A communication process in which data is collected, recorded, processed, and distributed for operational purposes.

*System, Real-Time:* A real-time system may be thought of as a communications-oriented, data-processing system capable of performing batch-processing functions while concurrently processing inquiries, messages, and data transfer, and generating responses in a time interval directly related to the operational requirements of the system. Amplified by *Real-Time.*

*Tape:* A ribbon of flexible material used as a storage medium, described by a qualifying adjective such as paper, magnetic, oiled, etc.

*Tape, Magnetic:* A tape or ribbon of any material impregnated or coated with magnetic or other material on which information may be placed in the form of magnetically polarized spots.

*Tape, Paper:* A strip of paper used to store or record information in the form of punched holes, partially-punched holes, and/or imprinting. See *Tape, Punched Paper* for amplification.

*Tape, Punched Paper:* Paper tape on which information has been (or may be) stored or recorded in the form of partially-punched (chadless) or fully-punched (chadded) holes. Each character of information is punched in binary representation in a column across the width of the tape. There are usually 5 to 8 punch positions (channels or levels) per column. The channels, or levels, run in rows parallel to the length of the tape. The punch positions are normally spaced ten to the inch.

*Tariff:* The published rate for a specific unit of equipment, facility, or type of service provided by a communications common carrier.

*Telecommunications:* The transmission of data over a distance, usually by electrical means. Contrasts with directly-connected equipment in close proximity.

*Telegraphy:* A system of communication for the transmission of graphic symbols, usually letters or numerals, by use of a signal code. Now includes the use of teletypewriters.

*Telemetry:* Transmission of data over great distances, especially remote control of apparatus and equipment.

*Teleprocessing (IBM):* 1) An IBM trademark referring to their equipment and services used in data communications. 2) A form of information handling in which a data processing system utilizes communications facilities.

*Teletype:* Trade mark of the Teletype Corporation, usually used in reference to a series of different types of teleprinter equipment such as transmitters, tape punches, reperforators, page printers, etc., used for communications systems.

*Teletypewriter Exchange Service (TWX):* An automatic, dialed teleprinter exchange switching service provided by the Bell Telephone System on a commercial toll basis. A monthly rental is charged for the terminal equipment and a toll charge with a 1-minute minimum for each connection made (for interstate calls). It can be used to dial any other TWX subscriber in the United States and transmits at either 60 or 100 words per minute, depending on the equipment used.

*TELEX:* 1) An automatic teleprinter exchange switching service provided by Western Union. 2) An international network of teletypewriter subscriber service.

*Terminal:* 1) A point in a system or communication network at which data can either enter or leave. 2) An input/output device capable of transmitting entries to and obtaining output from the system of which it is a part.

*Text:* The part of a message that contains the substantive information to be conveyed. Sometimes called the body of the message.

*Throughput:* 1) The rate at which information can be accurately delivered when averaged over a long period of time. 2) The time required to perform an operation from the time it begins until the time it is successfully completed.

*Time, Propagation:* The time required for a signal pulse to travel from one point to another. Signals travel at different speeds through different media, as the following rules of thumb indicate:

| *Medium* | *miles/sec* | *ms/mile* |
|---|---|---|
| Carrier: | 100,000 | 0.01 |
| Wire: | 14,000 | 0.0714 |
| Loc. Chan.: | 8000 | 0.125 |

Equipment: about 1.2 ms per line.
Repeaters: about 60 μs each.
Impedance matching coils: 20 μs.

Also see *Delay, Absolute.*

*Time, Turnaround:* 1) The time required to reverse the echo-suppressors on a switched telephone circuit. 2) The time required for a system to transfer from the receive mode to the transmit mode, or vice versa.

*Tone:* The audible result of a frequency within the audio range (approximately 20 to 20,000 Hz).

*Total, Batch:* Same as *Total, Hash.*

*Total, Hash:* A sum formed for error-checking purposes by adding fields that are not normally related by unit of measure, e.g., a total of invoice serial numbers.

*Transceiver:* A device that both transmits and receives data, often simultaneously.

*Transmission, Carrier:* Transmission in which the transmitted electric wave results from the modulation of a single-frequency wave by a modulating (information-carrying) wave.

*Transparent:* The property of being insensitive to the meaning of a code being manipulated. An example would be a paper-tape transmitter capable of transmitting any code submitted to it. If a device interprets and reacts to coded information it is handling, it is said to be code sensitive (not transparent).

*Trunk:* A circuit between two telephone exchanges or switching centers, or from an exchange to a customer's switchboard.

*Type, Channel:* Simplex, half-duplex, and duplex circuits are types of channels. The names indicate only the directional capability of the channel. Also see *Channel, Class of* and *Channel, Grade of.*

*Unit:* 1) A basic element. 2) A device having a special function.

*Unit, Central Processing:* 1) The main portion of a computer complex. It contains the primary storage, arithmetic and logic units, and the control unit. 2) All that portion of a computer exclusive of the input, output, peripheral, and secondary storage units. Same as CPU and *Frame, Main.*

*Utility, Computer:* A service that provides computational ability. A time-shared computer system. Certain data and programs are shared by all users of the service; other data and programs, because of proprietary nature, have restricted access. Computer utilities are generally accessed by means of data communication subsystems.

*V-Answerback:* The response to a transmitter start code (TSC) of a polled station in a teletypewriter switched network if the station has no traffic to send, or the response to a call-directing code (CDC) of a station that is prepared to receive traffic.

*Voice-Grade Channel:* See *Channel, Voice-Grade.*

*XMT:* Transmit.

*XMTR:* Transmitter.

# Index

ABERYSTWYTH U.C.W. LIBRARY